The Battle of Sacramento

Forrest's First Fight

A Skirmish of Future Generals

by

John K. 'Ken' Ward

The Battle of Sacramento

Forrest's First Fight

A Skirmish of Future Generals

BY

John K. "Ken" Ward

HERITAGE BOOKS
2012

HERITAGE BOOKS
AN IMPRINT OF HERITAGE BOOKS, INC.

Books, CDs, and more—Worldwide

For our listing of thousands of titles see our website
at
www.HeritageBooks.com

Published 2012 by
HERITAGE BOOKS, INC.
Publishing Division
100 Railroad Ave. #104
Westminster, Maryland 21157

International Standard Book Numbers
Paperbound: 978-0-7884-5405-9
Clothbound: 978-0-7884-9326-3

Thanks To

Richard Alexander, San Francisco, Calif.

John Anderson, Preservation Officer, Archives & Information Services Division, Texas State Library and Archives Commission, Austin, Tex.

Cora Bennett, McLean Co., Ky.

Elizabeth Cox, McLean Co., Ky.

Kentucky Historical Society Library, Frankfort, Ky.

Kendall Milton, Curator, Texas Heritage Museum, Hillsboro, Tex.

John W. Muster III, McLean Co., Ky.

Lt. Doug Swiggett, University of Kentucky, Lexington, Ky., Police Department and nine-year U.S. Army veteran.

Tennessee State Library and Archives, Nashville, Tenn.

Ann Vance Todd, Frankfort, Ky.

Universtiy of Kentucky Libraries, especially Archives, Lexington, Ky.

Landon Wills, McLean Co. and Frankfort, Ky., and the Landon Wills Estate.

and expecially my parents, the late John Henry Ward and the late Ollie Kittinger Ward Cummings, who taught me to love history; and my brother, Hugh Ward, my critic, editor and friend.

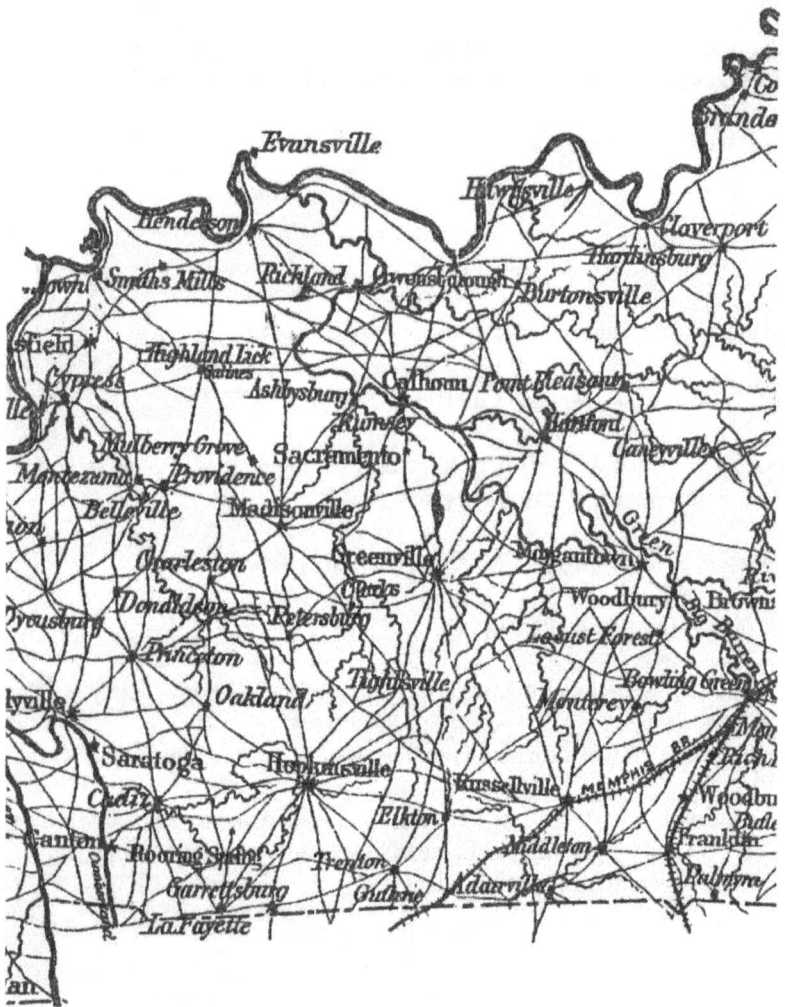

Area of the movements surrounding the battle and the battle itself. From a map accompanying the official records.

Area of the movements surrounding the battle and the battle itself.
From a 1958 U.S. Geological Survey map.

Contents

Battle of Sacramento

Forrest's First Cavalry Fight,   A Skirmish of Future Generals

(The military action at Sacramento could be viewed as a minor affair, of little or no consequence. From a military perspective, however, the basic elements of this skirmish were the same as those every soldier has faced and endured from before the Mycenaean Invasion of prehistoric Greece, c. 2000 B.C., to the current fighting in Iraq and Afghanistan. This tiny slice of combat, at a small village in west-central Kentucky, is actually a microcosm, not only of the Civil War, but of all wars. From another perspective, it is probable that no other skirmish this small had so many participants who achieved high-ranking positions, both during and after the war.[1])

(Note: Confederates are also called Southerners, Rebels or grey coats; Union troops are also called Federals, Yankees and blue coats.)

Initial developments leading to the cavalry engagement at Sacramento, Ky., on December 28, 1861, occurred during the previous month. In November 1861, Confederate Lt. Col. Nathan Bedford Forrest's Tennessee cavalry battalion was assigned to Hopkinsville, Ky., at that time a major outpost on the Confederate defense line in Kentucky. The unit had been operating out of Forts Henry and Donelson, Tenn. From Hopkinsville, the battalion engaged in reconnaissance patrols between the Green and Cumberland Rivers. Here they attempted to obtain supplies for the Confederate forces, harrass Federals and gain information about Union strength and movements. Brig. Gen. Charles Clark served as Forrest's immediate commander.[2]

Nathan Bedford Forrest—he went by Bedford—grew up poor and hard on the frontier of southwestern Tennessee and northwestern Mississippi. Ambitious and hard working, he became wealthy as a horse, cattle and slave trader and land dealer with extensive planting interests of his own. His wealth at the beginning of the war, Forrest later estimated, stood at $1.5-million.

> "And if both sides nourished the boldest hopes and put forth their utmost strength for the war, this was only natural. Zeal is always at its height at the commencement of an undertaking; and on this particular occasion Peloponnese and Athens were both full of young men whose inexperience made them eager to take up arms..."
>
> Thucydides
> *The Peloponnesian War*
> c.400 B.C., rep. Rex Warner, trans., Baltimore, Md.1974.
> (Book II, No. 8) p. 91

Bedford, now making his home in Memphis, initially opposed secession but when Tennessee seceded joined the fight on the side of his state.

Forrest first joined Capt. Josiah White's Tennessee Mounted Rifle Company as a private. But influential citizens convinced Tennessee Gov. Isham G. Harris that Forrest's talents were being wasted. The governor ordered Forrest back to Memphis and authorized him to raise a volunteer battalion of cavalry. Realizing the Confederacy was short of arms and

photo courtesy Library of Congress

Then Lt. Col. Nathan Bedford Forrest led the Confederate squadron at Sacramento.

equipment, Forrest traveled to Louisville, Ky., and purchased, with his own money, 500 Colt navy pistols and 100 saddles and other horse equipment. He successfully brought them off to Tennessee, along with his first company, Capt. Frank Overton's Boone Rangers, from Meade and Breckinridge Counties, Ky.[3]

Forrest's battalion was organized at Memphis, Tenn., in October 1861. At that time it was composed of Capt. J.F. Overton's Company (A), Capt. W.C. Bacot's Company (B), Capt. Charles May's Company (C), Capt. Nicholas C. Gould's Company (D), Capt. A.S. Truitt's Company (E), D.C. Kelley's Kelley Troopers (F) and Capt. M.D. Logan's Company (G), altogether 650 men. This Tennessee cavalry battalion then consisted of one Texas company(D), two Kentucky companies (A and G), four Alabama companies (B, E, F and H) and one Tennessee company (C). Forrest served as commander with the rank of lieutenant colonel, with D.C. Kelley as major.[4]

As soon as organization was complete, Forrest's squadron was ordered

to Ft. Donelson, Tenn., on the Cumberland River. From here they were to patrol the surrounding area. But on November 14, Forrest wrote General Albert Sidney Johnston, Confederate commander of the area west of the Appalachians, that he found the country impracticable for cavalry, and with scant subsistence. This had necessitated his splitting his small command. He requested that he be allowed to unite his command under Brig. Gen. Lloyd Tilghman at Hopkinsville. The move was approved but by that time General Tilghman was no longer in command at Hopkinsville. Brig. Gen. Charles Clark had been placed in charge of that post.[5]

During the late December scouting expedition, Forrest took with him, from his own battalion, detachments from Companies A, C and D, under 1st Lt. John Cruther and Captains May and Gould; and details from Companies E, F and G as a detachment under Maj. David C. Kelley. In addition to these elements of his battalion, Forrest had with him at Sacramento a 25-man scouting unit under Capt. Charles Edward 'Ned' Meriwether, of Todd Co., Ky., and a 40-man detachment from Capt. W.S. McLemore's Company F, 8th Tennessee Cavalry Battalion, under the battalion commander, Lt. Col. James W. Starnes.[6]

> "The earliest records of mounted combatants as a distinct military organization date far back in the history of Asia. Diodorus, a Greek historian of the first century B.C., states that Osymandias, who lived before the Trojan War (1230-1190 B.C.), led 20,000 mounted men against the rebels in Bactriana. In the first Messenian War, 743 B.C., Lycurgus formed his cavalry in divisions. In the battle of Arbela, 331 B.C., Alexander, leading 7,000 of the Macedonian cavalry, dashed into a gap of the Persian army, and by this brilliant feat of daring and skill utterly routed the enemy, thereby establishing cavalry as a decisive tactical factor."
>
> "The Tactical Evolution Of Cavalry,"
> Cavalry Combat
> U.S. Army Cavalry School, 1937
> pp. 1-7, p. 1

Capt. Edward 'Ned' Meriwether and another man identified only as Childers recruited a unit from the Todd County, Ky., and the adjacent Tennessee area. Some differences arose in the company and Childers and 20 of the men left to join the Confederate Army in the East. The remaining men, under Meriwether, went to Camp Boone, Tenn. Here they were mustered into an infantry regiment but with the understanding that when Meriwether recruited enough men to form a company they should be released. When the Confederates arrived in Hopkinsville, Meriwether demanded the agreement be carried out. After some difficulty the men were released from the infantry.

> "In the United States, cavalry became a force to be reckoned with during the Civil War. Cavalry activity often closely conformed to traditional experience. For instance, mounted riders gathered information about opposing forces, destroyed bridges and engaged in other harassing actions, and provided general messenger and mobile guard details. The dash and cunning of a few imaginative cavalry leaders also gave a new expanded dimension to cavalry service. Their names rang like rattling sabers across the pages of history: Jeb Stuart, Custer, Sheridan, and Forrest. They grouped cavalrymen into independent formations equally capable of executing swift attacks or decisive delaying actions and sent reinforced units of riders patrolling through the broken eastern woodlands to cover the flanks of entire armies marching on campaign."
>
> Shelby L. Stanton,
> *Anatomy of A Division:*
> *1st Cav in Vietnam*, Novato, Calif., 1987, p. 5

Forming a cavalry unit, the captain's troopers called themselves the "Meriwether Independent Rangers." On October 29, 1861, General Tilghman reported from Hopkinsville: "Of cavalry we have nothing to count on, save Captain Meriwether's company of untutored recruits." John Headley wrote that when he arrived in Hopkinsville to join the Confederate Army, Captain Meriwether's cavalry company was stationed there. Ed Porter Thompson, in his history of the Orphan Brigade, states that Meriwether's was one of two companies being recruited for the Confederate 1st Kentucky Cavalry Regiment. Meriwether's unit was expected to be Company I. An officer of the other company, planned Company K, reported: "With Meriwether's (afterward Williams's) company," his unit "operated as a Kentucky squadron under command of Colonel Forrest..." (In several accounts, especially early newspaper accounts of the Sacramento battle, Captain Meriwether's name is given as H. Clay Meriwether, of Louisville. Oddly enough, there was an H. Clay Meriwether who joined the Confederate forces. He was later captain of Company H, Johnson's 10th Kentucky Partisan Rangers, later 10th Kentucky Cavalry, and enlisted in Daviess Co., Ky., September 5, 1862. A writer from the 11th Kentucky Infantry, at Calhoun, stated both Meriwethers were killed at Sacramento.)[7]

A Mexican War veteran and former physician, James W. Starnes claimed to have withdrawn from medicine to manage his plantation on the Halpeth River. He gave up his practice, however, after he had been unable to save his first two children when they became gravely ill. Starnes organized the "Williamson County Cavalry," also known as

photo from *Dover Publications*

A pontoon bridge, like this one across the Rappahannock in 1863, connected Calhoun and Rumsey in late 1861.

Captain Starnes' Company Tennessee Volunteers, on October 30, 1861, at Nashville, Tenn. William S. McLemore, former Williamson Co. court clerk, served as deputy commander. On December 11, with the addition of five more companies (A thru E) the 8th Tennessee Cavalry Battalion was organized at Camp Cheatham, Tenn. Starnes was promoted to lieutenant colonel, commanding, with Maj. Ewing A. Wilson as deputy. McLemore took command of the Williamson County Cavalry, now Company F. Stationed at Russellville, part of the 8th Cavalry Battalion rode out of that town early on the morning of December 28. They overtook their fellow Confederates at Greenville.[8]

On Thursday, December 26, Forrest set out on a reconnaissance north toward Green River. Four miles north of Hopkinsville, Colonel Forrest, with Companies A, C and D from his battalion, left the Greenville Road and moved toward Rochester. Captain Meriwether and the remaining details under Major Kelley rode on toward Greenville. The night of December 26 Major Kelley's detachment camped at Pond River. The next day they rode on to Greenville.[9]

Finding no enemy soldiers in the Greenville area, Major Kelley went into camp that night about half-a-mile north of town. The people of Greenville sent out an invitation to supper and the Confederate troopers enjoyed a "royal feast" at the courthouse. Just as Kelley's soldiers finished their meals, Colonel Starnes rode up with news that he had encountered and engaged a Union picket outpost at South Carrollton and suggested a Federal scouting party might still be operating in the area. Major Kelley immediately ordered his men back to camp and set out pickets. Colonel Starnes then moved on to his camp at Russellville but the next day returned to join Forrest.[10]

Meanwhile Colonel Forrest, finding no Federals in the Rochester area, joined Major Kelley on the morning of Saturday, December 28. With this juncture, and the return of Starnes with 40 of his men, Forrest numbered his command at about 300. Having encountered no sign of Federal forces, Forrest decided to move on toward Rumsey. Rumsey was a river town directly across Green River from Calhoun, the head-quarters of a Union division.[11]

At about the same time Forrest moved his command into Hopkinsville, Union Brig. Gen. Thomas L. Crittenden was moving into his new headquarters at Calhoun. Thomas Crittenden was the son of John J. Crittenden, a leading Kentucky statesman and author of the Crittenden Compro-mise, an attempt to head off civil

photo from Adam R Johnson's *Partisan Rangers*

Robert Martin served as a scout for Confederate Colonel Forrest at Sacramento

war. The Crittendens became one of the more noted divided families in Kentucky. Thomas' older brother, George, was a Confederate general.[12]

Thomas Crittenden was appointed brigadier general of volunteers on September 27, 1861. Early the following month he was ordered to take command of U.S. forces at Owensboro and Henderson. He moved his headquarters to Calhoun on Green River in November. On December 3 his command was designated 5th Division, Army of the Ohio. It was Crittenden's division Forrest was attempting to scout.[13]

As Forrest moved toward Rumsey that late December day, his Ken-tucky-born scouts, Robert Martin, from Greenville, Ky., and Adam R. Johnson, originally from Henderson, Ky., but now of Texas, were scouting far ahead. Both scouts were familiar with the region and had gone as far as Rumsey itself. At Rumsey they found that Federals had constructed a pontoon bridge connecting the Union camp at Calhoun, on the north side of Green River, and Rumsey, on the south side.[14]

Martin and Johnson were both exceptional individuals. Robert Maxwell 'Bob' Martin was born in Muhlenberg Co., Ky., January 10, 1840. Bob's six siblings—three brothers, three sisters—and his parents and some other relations were all Union sympathizers. Two brothers, Lt. Templeton

B. Martin, 11th Kentucky Infantry, and Lt. James H. Martin, 35th Kentucky Mounted Infantry, served as Union officers. Only Bob elected to side with the South. He early joined Bedford Forrest who quickly recognized Martin's ability as a scout. Adam Johnson described Martin as "over six feet in height, and although somewhat slender, possessed a well-knit figure... There was nothing in his outward appearance that bespoke the wonderful courage and daring that he continually displayed during the ensuirng years of the war."[15]

Born in Henderson, Ky., February 8, 1834, Adam Rankin Johnson spent much of his youth hunting in the then still wild areas of Henderson County. He gained a reputation as one of the best hunter in the region. "In this outdoor life," Johnson later wrote, "I acquired health, strength and activity and the habits of close observation and prompt action." At age 20 he moved to western Texas. Here he hunted, surveyed and fought Indians. He bought and successfully ran the Overland Mail's Staked Plains Station, "the most dangerous point on the line." Throughout these years he learned additional lessons in scouting, hunting and fighting that

stood him in good stead during the Civil War. With the commencement of war, he returned to Kentucky, leaving a young bride in Texas. At Hopkinsville he met Colonel Forrest, liked what he saw and volunteered as a scout. Forrest said he could use him and said he had one good scout, Bob Martin. "If you can equal him as a scout," Forrest told Johnson, "I will have a good team."[16]

While the scouts spent the night of December 27-28 attempting to ascertain Crittenden's intentions, Forrest, early on the morning of December 28, pulled out of

photo from Adam R Johnson's *Partisan Rangers*

Confederate scout Adam R. Johnson first reported the Federals to Colonel Forrest.

Greenville and headed north. Many
of his soldiers encountered a
Southern sympathizer in the over-
seer of the Weir farm just north of
Greenville. He gave the troopers a
late breakfast of milk and honey and
filled their knapsacks with the best
from the smokehouse.[17]

As the main Confederate force
moved toward Sacramento, they
found the road heavy with deep mud
and rough with slushy ice. (Cartoon-
ist, writer and World War II veteran
Bill Mauldin felt that mud was "a
curse which seems to save itself for

> "There is an infinite amount of
> hardship and drudgery con-
> nected with service in the ranks
> of any cavalry. It is necessary,
> therefore, to have not only
> ability to ride and intelligence
> to reconnoiter, but capacity in
> both man and horse to sustain
> long-continued exertion of the
> most arduous character. If
> either man or horse becomes
> exhausted or loses spirit, the
> effect is soon felt by the other."
>
> Gen. William H. Carter
> *The U.S. Cavalry Horse* , 1895,
> rep. Guilford, Conn. 2003, p. 5.

war." And Canadian Lt. Gen. E.L.M. Burns, veteran of World Wars I
and II, called mud "the soldier's ancient enemy.") It had started to rain as
the troopers left Hopkinsville and then began to freeze. The ground froze
during the night of December 27-28—behind the clouds a thin sliver of
moon hid, only three days from a new moon— but began to thaw as the
morning wore on. A cold but gentle rain again fell. The road was "exces-
sively severe on the animals." An advance guard under Captains
Meriwether and McLemore rode well to the front. Forrest rode at the
head of the main body, with Major Kelley with the center of the column.
Colonel Starnes brought up the rear.[18]

Just south of Sacramento Adam Johnson made contact with Forrest.
Johnson and his fellow scout had discovered a Union patrol south of the
river. Martin was keeping watch on the Federal force. As soon as the
scout reported, Forrest's troopers "were ordered to a rapid pace, and as
the news of the proximity of the enemy ran down the column it was
impossible to repress jubilant and defiant shouts, which reached the height
of enthusiasm as the women from the houses waved us forward."
Bravery can be as contagious as fear. The Union troops were part of a
scouting force from the 3d Kentucky Cavalry based at Calhoun.[19]

In early fall 1861 Kentucky Second District Congressman James S.
Jackson received authorization from Union Pres. Abraham Lincoln to
raise a regiment of cavalry for war service. He promptly resigned his
seat in Washington. Born in Fayette County, Ky., Jackson, a lawyer, had
served briefly in the Mexican War as a lieutenant in the 1st Kentucky
Cavalry. During this time he engaged in a duel with Thomas F. Marshall,
of the same regiment. Although the affair ended harmlessly, Jackson

feared he would be court martialed so he resigned and returned home. Prior to 1859 he moved from Greenup County to Hopkinsville.[20]

On September 6, 1861, Jackson published a call for volunteers for his unit. "I intend," he wrote, "to make this regiment in all respects equal to the best drilled and disciplined corps in the regular army." Captains and lieutenants were to be elected by the men of the individual companies. Companies were to remain at organization points until ordered to camp. Captains were to report to Colonel

photo courtsey Library of Congress

Then Maj. Eli Huston Murray led the Federal squadron at Sacramento.

Jackson at the Galt House in Louisville. One of the first companies organized was in Breckinridge Co., Ky., by Eli H. Murray who was elected captain. "He was a tall, handsome young man," Maj. Alfred Pirtle described him, "and his fine character and position, united with the fire and dash of youth, made him an ideal soldier, and it developed that he was peculiarly gifted for active field service." On November 26, 1861, even before the unit was mustered into U.S. service, Murray was promoted to major.[21]

On October 6, the first units of what became the 3d Kentucky Cavalry Regiment arrived at Owensboro, Ky., aboard the steamer *Hettie Gilmore*. Here and at Calhoun they completed their organization. Most of the men came from the Green River region, west of the Louisville & Nashville Railroad, with many coming from Christian and adjacent counties. Although not yet officially mustered into active service, detachments began active scouting immediately after arriving at Owensboro, sometimes engaging Confederate cavalry detachments. At Calhoun, on December 13, 1861, Regular Army Maj. W.W. Sidell officially mustered the 3d Kentucky Cavalry into U.S. service.[22]

Until early December 1861 the infantry based in Calhoun undertook most of the Federal division's patrol work. But that ended. "(S)couting is being done away with by infantry, the cavalry being relied on most for that kind of business," Lt. Col. James M. Shanklin, deputy commander of the Union 42d Indiana Infantry, wrote his wife on December 13, 1861.[23]

The 3d Kentucky detachment sent out in late December was dispatched after General Crittenden received information of a Confederate cavalry detachment moving through the country just south of Rumsey. Another report stated that 80 Southern soldiers were in South Carrollton. As noted, Confederate Colonel Starnes, with a detachment from his 8th Tennessee Cavalry Battalion, had been on a scout to South Carrollton on December 26. He reported engaging a Federal picket and killing three.

The Union patrol had been dispatched Friday night, December 27, leaving camp sometime between nine o'clock and midnight (reported times differ). The force consisted of two squadrons, one commanded by Maj. W.S.D. Megowan, the other by Maj. Eli H. Murray, under overall command of Major Murray. Both detachments went to South Carrollton, arriving there early Saturday morning, December 28. Finding no Confederates nor hearing of any, they split and headed back toward Calhoun. On the way back the squadron commanded by Major Megowan's force headed toward Madisonville. They went as far as "Fishburg" (Frostburg, on Pond River?) and then swung back to Rumsey and across the river to the camp at Calhoun.

photo from Otto A. Rothert's A History of Muhlenberg County.

Some Federal troopers were watering their horses at Garst's Pond when the Confederates attacked.

Murray's force was composed of elements of Companies A, B, C and D, 3d Kentucky Cavalry, and numbered 168 men. According to one report only 30 had Enfield rifles, the rest only sabers and pistols. This was the detachment reported to Forrest. (As a side note, on December 30, 1861, General Crittenden, in his report of that date, admitted the 3d Kentucky troopers were "badly armed, and, what is worse, have no confidence in their pistols." Another report states that the 26th Kentucky Infantry and 3d Kentucky Cavalry together "counted barely two hundred firearms, and those were inferior." Even earlier, on October 30, 1861, Brig. Gen. W.T. Sherman, Union commander in Kentucky, had gone so far as to suggest arming the 3d Kentucky with lances if Crittenden couldn't find enough firearms for the troopers.)[24]

A few miles south of Sacramento this part of the patrol paused to rest. New to war and not being alert, the Federals were scattered along the road for several hundred yards. Some were watering their horses in Garst's Pond. They were completely unaware of the nearby Confederate cavalry. The Southerners, however, were well informed. Not only Johnson had reported the presence of the Federals but so had a local Southern sympathizer.[25]

"A beautiful young lady, smiling, with untied tresses floating in the breeze, met the column just before our advance guard came up with the rear of the enemy," Forrest reported, "infusing nerve into my arm and kindling knightly chivalry within my heart." Adam Johnson remembered that she rode bareback and dashed up "frantically waving her hat, while her long hair was flying in the wind like a pennant, and her cheeks were afire with excitement."[26]

This beautiful young lady was Mollie S. Morehead. Her father, a Southern sympathizer, lived near Sacramento. Mollie and her sister, Bettie, had been on an errand and saw the Union soldiers. According to Mollie, she and her sister had ridden through the Federal force. The sister rode on to warn her father and brothers. Mollie rode to Colonel Forrest.

> "Finally, I believe strongly in women taking part in their country's defence, not because I am a women's libber but because I feel we have to share the responsibility. How can women send their husbands, their brothers and their sons off to fight, and just sit at home? War is terrible. I hate it. But if we are going to have it, I think it is the responsibility of all of us."
>
> Yaffa,
> female Israeli Palmach fighter quoted in Shelley Saywell's *Women in War*, N.Y., 1985, p. 311

Forrest had some difficulty keeping the enthusiastic young lady from riding into battle with him. She exemplified the description of Matilda,, leader of Christian armies in the 11th century, that "courage and valour in mankind is not indeed a matter of sex, but of heart and spirit."[27]

(Mollie's display of bravery—and audacity—is one of the more interesting and delightful aspects of the Battle of Sacramento. But that this incident stands out tells more about ourselves, perhaps, than Mollie Morehead. In 1985 Shelley Saywell published her *Women In War,* in which she researched, usually by interviewing female veterans, women who served in wars from World War II to El Salvador. More than 100 years after the American Civil War she could still write: "The things I expected to hear and perhaps, in the beginning, wanted to hear were not always said... I wanted to hear that women are innately more pacifist than men, but I learned that they can be every bit as determined in their willingness to kill and die for their beliefs." But as Jean V. Berlin writes, in the introduction to *Women in the Civil War:* "Scholars generally agree that the Civil War was as important a watershed in the history of American women as it was in the history of the nation."[28])

> "I do not think it is too arbitrary to assert that in all the history of war, cold and snow have inflicted less misery and hardship on the soldier than has mud."
>
> Canadian Lt. Gen. E.L.M. Burns, quoted in C.E. Wood's *Mud: A Military History,* p. 77

By the time the young lady contacted Forrest, discipline in the Southern force had been completely lost. Adam Johnson viewed "this disorderly mass of men galloping pell mell at break neck speed" with a "fearful, sickening dread." Nor was Johnson the only Confederate present to describe this bedlam. "Some of the troopers who had fast horses went to the front," James Hammer, one of Forrest's troopers, wrote immediately afterward. "I never saw so much excitement in my life as on this occasion. Some of the horses fell down, others gave out, some ran away, among them my little black. He ran right up to the head of the column, and would have ran still further if I had not run by Maj. Kelley and he caught the bridle-rein and stopped him..."[29]

Trooper H.T. Gray was in the rear company of the Confederate column. The mud flying from the horses hooves to their front soon had the rear company covered with mud. "(O)ur boys began cursing the two companies ahead of us, whom we thought were riding too slow," Gray later recalled. They were discovering, as soldiers have discovered over the millenia, that war tends to be basically a very dirty business. Colonel Starnes, in charge of this rear company, told the men to pass the forward companies by separating columns and riding on the right and left of those in front.[30]

After receiving Miss Morehead's report of the near location of the Union patrol, Forrest sent Johnson ahead to exactly locate the enemy. Moving to a high point on one side of the road not far in advance of the Confederate force, the scout found the Federal cavalry just over the crest of the hill. Johnson immediately returned and reported to Forrest. As soon as Forrest, still riding, heard Johnson's report, he galloped to the top of the hill from which Johnson had seen the Federal troops.

When the Confederates were seen on the hilltop the Union

photo from *A Comprehensive History of Texas,* Vol. 2; courtesy Texas State Library and Archives Commission

Then Confederate Capt. Nicholas C. Gould led a company of Texans at the Battle of Sacramento.

cavalrymen appeared uncertain as to their identity. They may have thought this might be Major Megowan's patrol who had returned. Forrest clarified the situation by taking a Maynard rifle from one of his men and firing at the Federals, "as a sort of gauge or challenge to battle." This was the opening shot of the skirmish at Sacramento. "The power to decide on an action," James L. Stokesbury pointed out, in an introduction to *Masters of the Art of Command,* "and the strength to see it through, are probably the most fundamental qualities of a great soldier."[31]

Before advancing farther Forrest sought to form his men into some kind of order. His captains attempted to bring their companies into line but most of the officers could collect only a few men around them. Captain Gould, leader of Company D composed of Texans, was the only officer to manage anything at all. Most of his men, having faster horses, were at the front anyway so they rallied quickly around their leader. This brought some semblance of order out of the chaos.[32]

With about 150 men Forrest charged the Union advance. "When that moment (when it is clear a fight is on) arrives, whether it is in a barroom fight or in a war," writer Ernest Hemingway knowingly stressed, "the thing to do is to hit your opponent the first punch and hit him as hard as

possible." And Alfred Thayer Mahan, American admiral and famous strategist, emphasized that the "true speed of war is not headlong precipitancy, but the unremitting energy which wastes no time." As for Forrest, it could be said of him, as it was said of Alexander the Great: "Boldness was his creed."

The Union troopers opened fire when Forrest got within 200 yards. At 80 yards Forrest ordered his men to fire. "After these rounds I found that my men were not up in sufficient numbers to pursue them with success," Forrest wrote in his report, "and as they showed signs of fight, I ordered the advance to fall back." Forrest at this time was engaging only Major Murray and 45 men of his command. "Major Murray behaved with great gallantry," General Crittenden reported on December 30, "and... repelled the charge, being seconded handsomely by about 45 men."[33]

With the arrival of the rest of his command, the Confederate leader dismounted part of his force armed with Sharps carbines and Maynard rifles. They were to maintain a steady fire into the Union troops. He then sent mounted troops to hit the Federal flanks, Major Kelley from the right flank, Colonel Starnes from the left. Forrest himself led the reminder of his force straight into the center of Murray's line. "Officers must not hesitate to lead," Gen. George Patton noted in his field notebook. "Before an attack is declared hopeless, the senior officer must lead an attack in person."[34]

"This (charge) was done with an animating shout, and all possible spirit and resolution, but in face of a sharp fire, under which the brave Captain Meriwether... fell, shot with two balls through the head, by the side of the commander." Meriwether had raced with Forrest at the head of the charge, "a Kentuckian who couldn't let Tennesseans outride him in his home state." Forrest felt that, had he lived, "Capt. Ned Meriwether would have been one of the great calvary officers of the Confederacy." Captain Meriwether was the first casualty of the battle.[35]

Before continuing, Forrest's tactic should be looked at more closely, for the tactic is important in Military Science. But Forrest had

> "In one respect a cavalry charge is very like ordinary life. So long as you are all right, firmly in your saddle, your horse in hand, and well armed, lots of enemies will give you a wide berth. But as soon as you have lost a stirrup, have a rein cut, have dropped your weapon, are wounded, or your horse is wounded, then is the moment when from all quarters enemies rush upon you..."
>
> Winston Churchill, "The Cavalry Charge at Omdurman," in Ernest Hemingway, ed., Men At War, 1942, rep. N.Y., 1955, pp. 813-821, p. 819

not studied Military Science. His later commander, Lt. Gen. Richard Taylor, once remarked that Forrest "employed the tactics of Frederick at Leuthen and Zorndorf, without even having heard these names."[36]

Sun Tzu, in the first known treatise on war, wrote of the quintessential general and his troops: "Without extorting their support the general obtains it; without inviting their affection he gains it; without demanding their trust he wins it." A somewhat later writer, in commenting on this passage wrote: "This refers to the troops of a general who nourishes them, who unites

photo from Jordan and Pryor, *The Campaigns of Lieut.-Gen. N.B. Forrest, and of Forrest's Cavalry*

Confederate Brig. Gen. James R. Chalmers discussed Forrest's tactics at Sacramento.

them in spirit, who husbands their strength, and who makes unfathomable plans." All of which Forrest seemed to know instinctively.[37]

"After all," Forrest opponent Union Gen. William T. Sherman said after the war, "I think Forrest was the most remarkable man our Civil War produced on either side. To my mind he was the most remarkable in many ways. In the first place, he was uneducated, while Jackson and Sheridan and other brilliant leaders were soldiers by profession. He had never read a military book in his life, knew nothing about tactics, could not even drill a company, but he had a genius for strategy which was original, and to me incomprehensible. There was no theory or art of war by which I could calculate with any degree of certainty what Forrest was up to. He seemed always to know what I was doing or intended to do, while I am free to confess I could never tell or form any satisfactory idea of what he was trying to accomplish."[38]

Brig. Gen. James R. Chalmers served under General Forrest during the last year and a half of the war. In 1879 he wrote that the battle at Sacramento "illustrated the military characteristics of the man." Forrest's military characteristics, as pointed out by General Chalmers, were:

"First, his reckless courage in making the attack—a rule which he invariably followed and which tended always to intimidate his adversary. Second, his quick dismounting of his men to fight, showing that he regarded horses mainly as a rapid means of transportation for his troops.

Third, his intuitive adoption of the flank attack, so successfully used by Alexander, Hannibal and Tamerlane—so demoralizing to an enemy even in an open field, and so much more so when made, as Forrest often did, under cover of woods which concealed the weakness of the attacking party. Fourth, his fierce and untiring pursuit, which so often changes retreat into rout and makes victory complete." During World War II the U.S. Army promoted a brief axiom: "Exploit success." Forrest applied that axiom instinctively.[39]

> "...when two opposing patrols in war clash in mortal combat, it ceases to be a military maneuver; it is a man-to-man struggle for self-preservation, a Battle Royal in which the only referee is one's conscience, then numbed by the specter of death. War is ghastly!"
>
> *Cavalry Combat*
> The United States Cavalry Association
> Harrisburg, Pa., 1937, p. 15

Crittenden wrote that Murray and his 45-man force "resisted the whole body of the enemy for ten minutes, and, from the accounts I have from many reliable witnesses, would have repulsed them, but at this critical moment some dastard unknown shouted 'Retreat to Sacramento!' Most of the men fled, of course, without stopping at Sacramento." Pierre Leulliette, a French paratrooper during the Algerian War (1954-1962), also found fear in combat. "What a mystery is fear," he later wrote, "at first sly and secret, then a beast devouring one's vitals."[40]

The 3d Kentucky's Capt. A.N. Davis, commanding Company D, later stated that the repulse of his detachment was due to the Federals' firearms. "... (O)ur troops only had about thirty guns

photo from *Photographic History of the Civil War*

Union cavalryman from the Western theater, between the Appalachian Mountains and the Mississippi, early in the war.

among them, and the remainder were armed with pistols furnished with cartridges, which were absolutely worthless." According to Captain Davis, his men would "ride up to within a few paces of the enemy, and take deliberate aim, and discharge their pistols, and that the Confederate soldiers would laugh in their faces at our shots." He was later told by the Confederates that when they were hit, the shot merely struck, then fell to the ground.[41]

War theorist Carl von Clausewitz wrote in the early 19th century that to introduce the principle of moderation into the theory of war would lead to "logical absurdity." War, Clausewitz wrote, "is an act of violence pushed to its utmost bounds." His words were seconded by Lord Thomas Macauley (1800-1859) who wrote that the "essence of war

photo from Otto A. Rothert's
A History of Muhlenberg County.

Although captured, Federal Capt. Arthur N. Davis received commendations for his bravery.

is violence and moderation in war is imbecility." Although Forrest had never heard of the classic writer on war nor the English historian, he definitely understood this principal of war. Once the Union troops broke Forrest never let up, striking their rear again and again. Charles W. Button described the pursuit as "like a fox-chase, the best-mounted men in front, regardless of order or organization." (Writer John Steinbeck, covering World War II, noted that most military operations quickly evolved into "disorganized insanity.")[42]

An advisor to King Agis of Sparta (427-400) suggested to the ruler, during a battle, that those enemy who were fleeing should be allowed to escape. The king replied: "Yet if we don't fight those who are fleeing out of cowardice, how shall we fight those brave enough to stand firm?"[43]

Confederate Colonel Forrest gave credit to a few Federal officers for attempting to halt the flight of the Union force. They were soon forced to join the fleeing cavalrymen. The Confederates did not catch up with the Federals until they reached Sacramento. At that point "there commenced a promiscuous saber slaughter of their rear." (At the beginning of another war, English writer Evelyn Waugh recorded in his diary, in October 1939, "They are saying, '...There are going to be no wholesale slaughters.' I ask how is victory possible except by wholesale slaughters?")[44]

The Federal troopers actually had more than the "promiscuous saber slaughter" to contend with as they tore through Sacramento. Several citizens of the village fired at the Union soldiers as they rode through. Some reports even erroneously stated that Union Capt. Albert Bacon was mortally wounded by a shot fired through the window of a house. A later report states the citizens fired on the Union troopers with shotguns filled with bird and squirrel shot. Many Federals at Calhoun strongly believed these citizens were also "implicated in the plot that resulted so disastrously to our soldiers." They believed that reports of Southern cavalry south of Green River were only a ruse to draw Federal cavalry into a trap. They thought the attack at Sacramento was the trap springing.[45]

> "Only by means of a relentless pursuit of the beaten enemy can the full fruits of victory be obtained. Pursuit of a decisively defeated enemy must be pushed to the utmost limit of the physical endurance of the troops and no opportunity given him to reorganize his forces and reconstitute his defense. The object of the pursuit is the annihilation of the hostile force... By the concentrated employment of every agency of destruction and terrorization at the disposal of the field forces, the shaken morale of the defeated enemy is converted into panic and the incipient dissolution of his organization is transformed into rout."
>
> *Field Service Regulations, U.S. Army, 1923* quoted in *Cavalry Combat,* U.S. Army Cavalry School, 1937, p. 328

The pursuit of the Union forces was continued at almost full speed for two miles beyond Sacramento. According to Forrest's report there were "bleeding and wounded strewn along the route." Trooper Button admitted that during the pursuit "no one stopped to take charge of a prisoner." The battle had now been pared to combat essentials: kill or be killed... or run. "You can say the words 'death and destruction' and they don't mean anything," famous war correspondent Martha Gellhorn wrote. "But they are awful words when you are looking at what they mean." But Napoleon would have approved the pursuit. "It is the business of cavalry to follow up the victory," Napoleon wrote, "and to prevent the beaten army from rallying."[46]

One Union casualty was Pvt. Calvin A. McCullough, Company A, serving as company quartermaster sergeant, wounded early in the fighting. A pistol ball struck him in the chest, perforating the right lung. He remained in the saddle but during the retreat he was again wounded. A saber thrust in the lower left side penetrated his bowels. "There could be no doubt that the sabre penetrated the bowels," the Union's Surgeon General's report states, "for there was a very copious discharge of fecal matter by the wound for several days." Incredibly, McCullough managed to remain in the saddle for another mile. Loss of blood finally brought him to a halt, too faint to remain on his horse. The wounded soldier was left at a farm house (owner not listed, unfortunately).[47]

Pvt. Rezin McDaniel (given as Rearson McDaniel in the Kentucky Adjutant General's Report, Reason McDaniel in the *History of Daviess County, Kentucky*, and Rezin in the 1850 census), Company C, was another Union soldier wounded at Sacramento. From Daviess Co., Ky., he was also captured in the Sacramento affair.[48]

"It was the first time I had seen the Colonel in the face of the enemy," Confederate Major Kelley, who would serve with Forrest throughout most of the war, wrote soon after the Sacramento fight, "and, when he rode up to me in the thick of the action, I could scarcely believe him to be the man I had known for several months. His face flushed till it bore a striking resemblance to a painted Indian warrior's, and his eyes, usually mild in their expression, were blazing with the intense glare of a panther's springing upon its prey. In fact, he looked as little like the Forrest of our mess-table as the storm of December resembles the quiet of June." (In a study of masters of the art of command, Martin Blumenson and James L. Stokesbury pointed out that "men seem to like, and even to desire, combat." As to why, they suggested that the reason "may be more fundamental than the philosophers think: in war everyone is young.")[49]

photo from John Allan Wyeth, *That Devil Forrest*, 1899

Then Maj. David C. Kelley, known as 'Forrest's Fighting Preacher,' served as Colonel Forrest's deputy commander at Sacramento.

"I am assured by officers and men," General Clark, Forrest's commander, later reported, " that throughout the entire engagement he was conspicuous for the most daring courage; always in advance of his command."[50]

Some U.S. officers finally managed to pull enough men together to slow and eventually halt the Confederate pursuit. During the final yards, the skirmish was marked by a series of hand-to-hand engagements. "(O)nce close-quarter combat begins all soldiers experience the same rush of emotions—exhilaration, fear and stress. And however sophisticated their weapons and equipment, when they see the whites of the enemy's eyes," historians of close quarters combat have pointed out, "the action is often nasty, brutish and short."[51]

Forrest, large himself, is said at one point to have encountered a Union soldier "as large as himself, muscular and powerful." While engaged in combat with this soldier, another Federal came up behind Forrest and was set to run his sword into the Confederate commander's back. A member of Forrest's command saw the action and shot this second Union trooper. Forrest won his saber duel and killed the first soldier. This victim of Forrest's saber may have been Corp. Isaac Mitchell, Company D. A half-inch short of six feet tall, he was a farmer from near Greenville. He had been born in Tennessee.[52]

Another Greenville native was definitely involved with Forrest. Union Pvt. John L. Williams, Company D, 3d Kentucky, had been unhorsed, so Forrest rode up and demanded Williams' surrender. According to Forrest's statement in Greenville that night, "Williams looked him straight in the eye, and drew back his pistol and threw it with great force striking (Forrest) on the breast and would have knocked him off his horse had he not been a large man; that immediately some of (Forrest's) men rushed up and began using sabers on Williams, but (Forrest) stopped them at once. He remarked that Williams was too brave a man to be butchered when overpowered." Some reports state Al Fowler, later a Confederate raider in the area, was the one who used his saber on Williams. Williams was taken as a prisoner to Greenville. There he was released on parole to his family. [53]

In his study of Persian Xerxes' 480 B.C. failed invasion of Greece, Ernle Bradford, World War II veteran and historian of Ancient Greece notes: "War, then as now, was an indescribable mixture of cruelty and violence coupled with admiration, in some cases, for the courage of an opponent."[54]

Other than Captain Meriwether, Pvt. William Hewlett 'Bill' Terry, of Captain McLemore's Company and a close neighbor of Colonel Starnes, was the only Confederate killed. Throughout the skirmish he had been conspicuous for gallantry. Terry was reportedly the Confederate who came to Forrest's aid when he was beset by two Federals. When Forrest had killed the large Union cavalryman, he turned back toward Private Terry and saw the private in a desperate struggle. He was battering a Federal trooper with the butt of his empty shotgun when he was killed with a saber thrust from Capt. Arthur N. Davis. Forrest rushed to assist Terry but was too late. [55]

General Crittenden described Captain Davis, commander of the 3d Cavalry's Company D, as also being "conspicuous in the fight for bravery." At one point in the fighting Davis came close to killing Forrest. Forrest stated in Greenville that "in the running fight Davis rushed up behind him, and that he would have received a fatal thrust had not Davis' horse fallen, for in the fall of the animal Davis' arm was injured and he rose and surrendered." Davis later stated, in an invalid pension application, that he was thrown from his horse "with such violence as to dislo-

photo from Otto A. Rothert's *A History of Muhlenberg County.*

The end of the battle occurred near this site. Station Baptist Church appears in the distance. Captain Bacon is said to have died near the tree on the right. Note the road looks much the same at the time of this photo as it did at the time of the fight, some 50 years earlier.

cate his right shoulder." At the time, he was "engaged in a hand to hand fight with the enemy."[56]

Another Union officer who received praise was also the highest ranking Union soldier killed. Capt. Albert G. Bacon, commander of Company C, was a 43-year-old Frankfort, Ky., native. He is credited with leading the attempt which finally halted the Confederate pursuit. He also engaged Forrest. Bacon narrowly missed Forrest with a saber stroke and Forrest's horse carried him ahead of Bacon. The Confederate leader then turned and shot the Union captain. Forrest ordered the wounded captain to surrender. Bacon attempted to continue the fight thus "obliging" Forrest "to run his saber through" the captain.[57]

photo from John Allan Wyeth, *That Devil Forrest*, 1899

Lt. Col. James W. Starnes led elements of his Confederate 8th Tennessee Cavalry Battalion at Sacramento.

One other Union officer was a casualty. 2d Lt. John L. Walters, Company B, was captured during the fighting.[58]

The mortal wounding of Captain Bacon brings up the point: where does fighting end and brutality begin? The authors of *The Whites of Their Eyes: Close-Quarter Combat* also look at this dilemma. "The deciding factor between what constitutes legitimate and illegitimate killing in wartime must surely be tactical necessity, but who decides?" they ask. "The aggressor, in the heat of the moment? Or those who come after, and cannot possibly know the full facts of any case? The question arises in combat of all sorts, but none provokes it more clearly than does close-quarter battle."[59]

Bacon's death ultimately led to the halting of the Confederate pursuit. The captain's horse and another riderless horse collided at full speed and knocked each other over at the bottom of a short, abrupt hillock. Forrest, again leading the pursuit, ran straight into the two horses. His horse added to the pileup but Forrest went flying over his horse's neck and landed 20 feet ahead of the heap of struggling horses. The three horses were added to almost immediately by a number of other horses and riders whose impetus carried them into the pile, effectively creating a road block.

Adam Johnson jumped his horse over the pile and saw his commander getting up. Forrest yelled for Johnson to get him a horse. Catching a free horse running down the road, he took it back to Forrest. Some of those running into the roadblock were Union cavalrymen already overtaken by Forrest and other Confederates. The Federals, now unhorsed, were captured.[60]

Colonel Starnes, ahead of the pileup, found himself leading the pursuit. The colonel, who could fire a pistol with either hand, attempted to engage a Union soldier. As the

photo courtesy National Archives

Then Brig. Gen. Thomas Crittenden commanded the Union 5th Division at Calhoun in December 1861.

Federal disengaged and renewed his retreat, Starnes threw his empty pistol at the departing opponent. The pistol struck the withdrawing soldier in the back, which "only helped to send him on his way."[61]

With most of the Union troops now out of sight and the camp at Calhoun being nearby, the Confederates abandoned further pursuit. Too, their

"The commander of a pursuit must be imbued with a resistless will to destroy, and this must be felt down to the last man. Without regard for their neighbors, or for their communications, everybody and everything must push on after the fleeing foe..."

Col. Herman Foertsch, "Pursuit," from *Art of Modern War*, quoted in *The Cavalry Journal*, Vol. 50, No. 5 (Sept.-Oct. 1941) p. 88

horses, wearied by a hard ride from Greenville, had then been pushed by the swift chase after the Federal troopers. Some horses had already dropped, as noted by Confederate trooper James Hammer. The men, too, mud-covered and very likely sweat-soaked and occasionally blood-splattered, surely now found the

> "... Valuable lessons are there to be learned from every battle, even the disasters or maybe especially the disasters..."
>
> Lt. Gen. (USA Ret.) Harold G., Moore and Joseph L. Galloway, *We are Soldiers Still* N.Y., 2008, p. 124

energy created by the adrenaline rush quickly ebbing away. The atmosphere probably reeked with the sharp bite of drifting whisps of black powder smoke, accumulated human and horse sweat, blood- and perspiration-wet clothing, perhaps some dust and mud tossed up by hooves still hung in the air. And maybe a tang of residual fear and excitement.[62]

Adam Johnson was ordered forward to reconnoiter. He gathered a few men and rode on to a high point where he could see the road. Finding it clear, he left the men there on guard and rode back to report to Colonel Forrest. Afterward he found Robert Martin. Johnson asked Martin what he had been doing. His fellow scout pointed to a horse and a belt full of pistols. Johnson asked him what happened to the Yank. "I left him over yonder in that strip of woods you see to the left of that road," Martin replied.[63]

Nor was Martin the only one to collect equipment from the Federal cavalry. According to a correspondence to the *Memphis* (Tenn.) *Daily Appeal,* the Confederates brought back to Hopkinsville at least 10 horses, 22 pistols, 18 sabres and 12 Enfield rifles. A report the day before states they brought back "as much of the spoils as they could carry, among

> "I had not felt as if I were at a war until now, but now I knew I was. It was a feeling I cannot describe;... There was certainly fear in that feeling, and courage. It made you walk carefully and listen hard and it lifted the heart."
>
> Correspondent Martha Gellhorn on her first taste of war, in 1937 during the Spanish Civil War Martha Gellhorn, *The Face of War,* rev. ed., N.Y., 1988, p. 15

which were guns, pistols, sabres, overcoats, etc." But they also "left a great quantity which they were unable to carry."[64]

In addition to Captain Bacon and Corporal Mitchell, the Union force lost an additional seven enlisted men killed. From Company A, Corp. Lafayette Phelps, and Pvts. James H. Phamps, William Ray and Uriah M. Underwood were killed. Company B lost Sgt. George T. Mays and Pvt. John W. Evans. Company

C lost Pvt. Robert Norris. All except William Ray were killed in action. Ray died less than a week afterward of wounds received in the skirmish.[65]

There were, according to General Crittenden's report of January 3, 1862, a "loss of 8 gallant soldiers, and 3 officers of uncommon bravery and soldierly qualities." The three officers were Capt. Bacon, killed; Captain Davis and Lieutenant Walters, captured. General Crittenden does not note that 1st Lt. Robert H. King, Company C, was slightly wounded. Nor does Crittenden mention the fact that at least 10 privates were captured and became prisoners of war in the South. (Unfortunately, Major Murray's report of the affair at Sacramento is missing. Murray, according to General Crittenden, included a report "of the conduct of all the officers under his command," as well as the action "of the entire command.") At least 12 Federals were wounded.[66]

Other Union casualties included Private Williams who was paroled at his home in Greenville. Pvt. Edward Baker, Company B, was also paroled at Greenville. A Greenville native, although confused about the Federal's rank, stated that Forrest "also brought a prisoner to Greenville whose name was, I think, Ed Baker, a lieutenant from Princeton, Kentucky. I saw this prisoner the next day at Reno's Hotel. He was badly wounded; shot in the legs, arms, and body, and was absolutely helpless..."[67]

Born in Coventry, England, Baker received gun shot wounds which perforated his body and limbs. As described by the surgeon at Louisville, Private Baker suffered: "gun shot wounds; viz, fracture of the upper third of the left fibula, the ball passing through the leg from without inward; wound of the right thigh (middle) the ball lodging in the femur; wound of right humerus, the ball lodging in the upper third of the bone, also fracture of the right scapula, the ball lodging in the spine of the scapula. This latter wound completely disables the right arm."[68]

A member of Forrest's command, writing from Hopkinsville on January 1, 1862, states the Confederates captured 15 prisoners, "brought thirteen to this place and left two at Greenville, unable to travel." The two were Williams and Edwards. The other 13 included Captain Davis, Lieutenant Walters and the 10 privates mentioned above. This leaves one unaccounted for. When Captain Davis wrote President Lincoln requesting exchange considerations for himself and 10 privates he may have been guessing at the number of enlisted men held captive. Or the

"There is no glory in war—only good men dying terrible deaths."

Vietnam War Lt. Larry Gwin, quoted in Harold G. Moore and Joseph L. Galloway, *We are Soldiers Still*, N.Y., 2008, p. 121

Confederate could have been off by one.[69]

There are no known statements from those captured at Sacramento describing their feelings at becoming prisoners of war. But there are other descriptions of this traumatic personal incident of war. John McElroy, a Union cavalryman in the 16th Illinois Cavalry, was captured at Jonesville, Va., near the Kentucky and Tennessee border, in December 1863. "We were overcome with rage and humiliation at being compelled to yield to an enemy whom we had hated so bitterly," Private McElroy wrote. "...(I)t seemed as if Fate could press to men's lips no cup with bitterer dregs in it than this." And Rev. T.J. Sheppard also later recalled that same calamity. "Never can I forget the mingled emotions of surprise, mortification and horror which I experienced when... I found myself hopelessly in the hand of the enemy," Sheppard wrote. "I thought I had considered every other chance of a soldier's fate when in the passion of patriotism I enlisted 'for three years or the war.' "[70]

> "What does a soldier feel when he is taken prisoner? First of all, that for him the war, and the danger of being wounded or even killed is over.
>
> "He then asks himself some worrisome questions: Where will he be going? How will they treat him? Will they torture him or even shoot him? He tries to master his rising fear and show courage, and he is concerned at first to suppress any thoughts about the implications of the lost war, or about his personal fate in the coming months or years."
>
> German Col. Hans von Luck,
> *Panzer Commander: Memoirs of Colonel Hans Von Luck,*
> 1989, rep. N.Y. 1991, p. 267

Forrest's report wildly overestimated the Union losses. His report to General Clark states: "There were killed on the field and mortally wounded, who have since died, about 65; wounded and taken prisoner, about 35, making their loss about 100." In forwarding Forrest's report, the general added, "The loss of the enemy, it will be seen, is estimated by Colonel Forrest at 65 killed and 35 wounded and prisoners, and from private and unofficial sources I learn that the number is not overestimated."[71]

(Forrest consistently includes a Captain Burges as a casualty with Captain Bacon, from the sabering onward. This was apparently a misidentification of an enlisted man. No Captain Burges appears in the roster of the 3d Kentucky Cavalry nor does General Crittenden include an officer named Burges among the casualties.)[72]

General Crittenden also overestimated enemy losses. He reported, "We do not know the extent of the enemy's loss. Meriwether (either a major or a lieutenant-colonel) was killed and certainly 4 men. The rebels took away three wagon loads of dead and wounded."[73]

When it comes to casualties at the Battle of Sacramento, as well as to the size of the opposing forces, overestimation was the norm. This is far from being unusual. As historian Ernle Bradford wrote, when discussing Persian King Xerxes' 480 B.C. campaign against the Greeks (with its most noted battle being the Spartan stand at Thermopylae): "Naturally enough, after the campaign was over, no Greek, whether soldier or sailor, was likely to reduce the number of ships and men that had come against him... It is a natural instinct of man to exaggerate, especially when comparing his prowess with that of an enemy."[74]

On January 7, 1862, the *Louisville* (Ky.) *Democrat* reprinted, from the *Columbia* (Mo.) *Statesman*, a "true statement" about "The Sacramento Affair." This "true statement," gave even higher Confederate losses. "Of the Rebels, thirty were killed and two six-mule wagon loads of wounded and thirteen prisoners." Forrest reported only Captain Meriwether and Private Terry killed and three privates slightly wounded. He failed to mention that Colonel Starnes received a saber cut on the head and one on each shoulder; none serious. (And where would they find two six-mule wagons in Sacramento in 1861?)

Neither Colonel Forrest nor General Crittenden mention captured Confederates. However, according to a *Louisville* (Ky.) *Daily Journal* report on January 11, 1862, there were five Southern prisoners taken. The item states that a detachment from the 3d Kentucky Cavalry had just arrived in Louisville with the prisoners "who were taken in the late skirmish at Sacramento." The five were identified as R.R. Robinson, Joseph M. Page, Reagan Hedrix, S. Rhodes and William Jett. These may have been civilians from Sacramento, although earlier reports stated the citizens were only forced to take the pledge of allegiance and then released. (A William Jett, Joseph Page, R.R. Robinson and Samuel Rhodes are listed in the 1860 McLean Co. census. No Reagan Hedrix or Hendrix appears. They could have

joined the Confederacy or been only McLean Co. citizens.)[75]

(Some later histories faired no better at counting casualties at Sacramento. In 1912 Benson J. Lossing's *A History of the Civil War* included a "Chronological Summary and Record of Every Engagement...compiled from the Official Records of the War Department." He reported Union losses as one killed and eight wounded and Confederate losses as 30 killed. Rossiter Johnson's earlier *Campfire and Battlefield,* did not explode the casualty figures, in fact, even admitted that it was "impossible to reconcile the accounts of the losses." He did, however, state that Major Murray's "small detachment of cavalry" was attacked by 700 Confederate cavalry under Colonel Forrest. He further wrote that the fighting lasted a half-hour, then the Union forces, "as their ammunition was exhausted, retreated.")[76]

In a January 7, 1862, letter to his family, James Hammer, a trooper in Forrest's cavalry, reported that one Southerner was shot by his own men. The Federals wore regular U.S. uniform overcoats; the Confederate, a Mexican War veteran, also wore a blue military coat. "We were all mixed up together," Hammer wrote, "and one of our men took him for a Yankee." Although Hammer believed his comrade would survive, he did consider the wounding a "sad misfortune as he is a good man."[77]

Such incidents as the wounding of the Confederate by his own men underscores the fact that "friendly fire" did not originate with the Vietnam War, when the phrase was coined. Numerous incidents of friends wounding or killing friends by accident occurred during the Civil War itself, the most notable being the mortal wounding of Lt. Gen. Thomas J. 'Stonewall" Jackson at Chancellorsville, Va., May 2, 1863, by his own men. Nor did such incidents suddenly first appear in the 1860s. They are known to have occurred throughout the history of recorded warfare.[78]

Colonel Forrest praised several of his men by name for their conduct during the fighting at Sacramento. These included Lt. Col. James Starnes; Maj. D.C. Kelley; Capt. N.C. Gould; Capt. Charles May; the fallen Capt. Ned Meriwether; Lt. John Crutcher, Co. A; Lieutenant (?) Nance; Lt. (?George L.) Cowan; Lt. Thomas W. Hampton, Company E, and especially the gallant conduct of Lieutenant (?) Bailey, of Captain Gould's company; Pvt. J.W. Ripley and Pvt. J.M. Luxton, Company C; and Pvt D.W. Johnson (Johnston in later newspaper report), Company G, "and, indeed, many others" who "did not come immediately under my own observation."[79]

In his report, Colonel Forrest writes that the Confederates found Union dead and wounded "in every direction." Those who could be moved were placed in wagons. The Confederates made Captain Bacon "as comfortable" as they could and "applied to the nearest farm house" to care for the Union officer, at least according to Forrest. But in depositions from Isaac Johnson and his wife, Elizabeth, after the Confederates departed they found Captain Bacon lying on the side of the road. He told them his watch had been taken. Another report states that not only was his watch taken but his pockets were rifled as well. A Southern correspondent admitted that each returning Confederate "had a trophy of some kind, some of which were valuable. ("It is said, and with some truth," writer John Steinbeck reported from Italy during World War II, "that while the Germans fight for world domination and the English for the defense of England, the Americans fight for souvenirs.")[80]

According to Pvt. D.W. Johnston (Johnson in Official Records), of Captain Logan's Company G, Forrest's battalion, the captain gave him the watch. Private Johnston wrote, on January 13, 1862, that he saw the wounded Union officer and dismounted to aid him. The Confederate wrote that he spread his blanket under Captain Bacon "in order to palliate his suffering as much as possible." Johnston then added that the captain took off his saber and belt and gave it to him, telling the private, "Take this and keep it and get the sabre, for it is an excellent blade." He further adds that the captain gave him his gold watch but did not say what to do with it. The captain added that he was a bachelor from Frankfort and wanted all his property to go to his sister. The Confederate ended by writing that the watch was still in his possession. "...(S)hould I ever return to that part of the state it is my intention to return it or have it returned to his sister."[81]

As soon as General Crittenden received word of the fight, he sent out a relief force. Commanded by 3d Kentucky Cavalry commander Colonel Jackson, the force consisted of about 500 men, 260 of them infantry from the 26th Kentucky Infantry Regiment. Capt. James M. Holmes, Company B, 3d Kentucky, led the advance to Sacramento. They pulled out of Calhoun at 10 p.m., December 28.[82]

Crittenden's instructions were "to gather up stragglers and the wounded, if there were any..." and "if the enemy were still in the vicinity to beat them up, but not to venture far in pursuit." Colonel Jackson's detachment remained in Sacramento that night and returned to camp at Calhoun late the following day. While at Sacramento Colonel Jackson buried the dead except for his friend, Captain Bacon. J.O. Shacklett, in a 1958 report, stated that the first persons to be buried in the Sacramento Cumberland Presbyterian cemetery were soldiers killed in the Sacramento battle. And one Union soldier was buried just back of the fence around Station Baptist Church cemetery. The body was later moved, as apparently were the bodies of the soldiers buried at the Cumberland Presbyterian cemetery.[83]

According to Onis Plain, a wounded Union soldier had fallen onto the road north of the Station Baptist Cemetery and was bleeding profusely. A Mr. Banks lived nearby. He hitched a team to a wagon and picked up the wounded soldier. Banks hauled the soldier to his residence and put him on the dining area floor. The Banks did what they could but the Federal died that night and was buried near the back of the cemetery. This was apparently the soldier buried just back of the fence around the cemetery. A darker spot remained on the dining floor for years, where the blood had soaked into the wood.[84]

By the time the Union force arrived at the scene of the battle, Forrest had already departed. He returned southward, arriving in Greenville late in the afternoon of December 28. That night he camped at Mt. Pisgah Church, near Pond River. Colonel Forrest and Major Kelley spent the night in a nearby house. From Mount Pisgah, the Confederates returned to Hopkinsville, arriving there on December 29.[85]

Apparently one member of Forrest's command was not with him at this time. This may have been one of his scouts, either Adam R. Johnson or Robert Martin. According to the story handed down in the Muster family of Calhoun, one of Forrest's troopers and his horse swam Green River below Calhoun that evening. He was familiar with the people of McLean County and made his way to the home of John W. Muster. At that time the Musters, who were strongly

photo from Otto A. Rothert's *A History of Muhlenberg County.*

The night after the battle, Colonel Forrest and his Confederate troops camped at Mt. Pisgah Church, near Pond River, in Muhlenberg Co.

pro-Southern, lived about five miles north of Calhoun. That night the Southern soldier was fed and given a place to sleep. The next morning he was again fed before he left.

During the battle the trooper had picked up a saber belonging to a Union cavalryman. The morning after the battle, while the young trooper waited for his breakfast, one of the young Muster boys used the captured sword as his "stick horse." When the soldier prepared to leave the boy began to cry because the trooper was taking his "horse" away. The soldier gave in and left the sword with the boy. The sword remains in the Muster family.[86]

The Musters' sword brings up an interesting point about the Battle of Sacramento. W.S. McLemore, in December 1861 a captain at Sacramento, still remembered the McLean Co. fight in 1900. McLemore served throughout the war, in the tough fighting engaged in by the Confederate cavalry in the West—the Civil War area between the Appalachian Mountains and the Mississippi River—and ended the war as a colonel. Surprisingly, the old soldier told an interviewer that Sacramento "was the only time (he) ever saw a hand-to-hand contest with sabers." So much for the movies multitude of Civil War cavalry saber fights.[87]

Apparently it was soon after Sacramento that Forrest voiced his annoyance with the traditional cavalry sabre which had only the end sharpened. When he struck with the blade, Bedford complained, the enemy just tumbled off and ran away. As far as that goes, note that Colonel Starnes received a saber cut on the head and one on each shoulder; but none were serious enough to warrant mention in Forrest's report. Eventually Bedford Forrest captured a saber that suited him and ground the blade to a razor sharp edge from hilt to tip.[88]

The combat at Sacramento was one of many small skirmishes and probing attacks which took place between the Battle of Bull Run, in July 1861, and Shiloh in April 1862 as both sides settled down to the business of war. The importance of these small fights was not so much in what they accomplished but in the experience gained by commanders and soldiers and the talents which began to emerge in the growing armies of the North and South. Kentucky saw her share of these on-the-job-training engagements. And Sacramento did contribute several intangibles to both sides.[89]

A well-grounded cliche states that soldiering is long hours of boredom broken by minutes of pure panic. Much of that boredom is made up of waiting. There is truly much waiting. Richard Holmes, in his study of the behavior of men in combat, found: "Much of a soldier's experience will, even in wartime, be of the everyday minutiae of military life rather than

> "... There is no tie of friendship so strong and lasting as that wrought by a common service between soldiers engaged in a common cause. Time and distance are powerless to sever such a tie, or to erase from memory the vivid recollections of dangers encountered and hardships endured..."
>
> Henry Lane Stone,
> "Reminiscences of Morgan's Men,"
> *Southern Bivouac*, Vol. 1, No. 11 (July. 1883), pp. 406-414, p. 407

the climax of battle, and many of the stresses which affect him will come as much from army life in general as from battle in particular." World War II war correspondent John Hersey described it thusly: "War is nine-tenths waiting—waiting in line for chow, waiting for promotion, waiting for mail, for an air raid, for dawn, for reinforcements, for orders, for the men in the front to move, for relief." But there is another kind of waiting.[90]

Before the first combat, there is the waiting of uncertainty. "How will I stack up? Will I prove myself a coward? Or will I stand tall, perform my job well, my duty faithfully?" This is the worst waiting. It is impossible to tell before the actual experience of battle whether you will be a good soldier or a failure. In a sense, whether a hero or a coward. Nothing in ordinary life or even in training quite prepares you for that actuality. "Combat is a unique experience," military historian Peter Maslowski has noted, "one that sets its participants apart from the overwhelmingly vast majority of the human race." Maslowski reinforces this by quoting two World War II riflemen after their first combat. They emphasized "nothing in civilian life or [military] training offered an experience remotely comparable." The first battle is truly a unique experience. (A Vietnam veteran called the trip into his first battle as "the wide river separating things known and secure from things unknown and insecure".)

But when it is over and you have passed that test, when you have, in the slang of the 1860s, "seen the elephant," there is a tremendous weight of uncertainty removed from the spirit. Writer and war correspondent John Steinbeck wrote of non-combat servicemen that "they lack only one thing to make them soldiers, enemy fire, and they will never be soldiers until they have it." A British platoon commander during the Falkland War (1982) put it another way. "The only real test of a man," he declared, "is when the firing starts." The troopers on both sides were now soldiers, had now gotten beyond that test. There will still be fear and anxiety—unless you are abnormal there will always be fear and anxiety. As William Breuer wrote: "Going into battle, there are two kinds of men who are *not* afraid—liars and bigger liars." But the knowledge that you leashed that fear once gives you confidence you can keep it in check in the future.[91]

The battle also usually helped the units involved achieve personalities of their own. Military units all tend to attain their own character. That personality can be cowardly, disrespective, even infamous. Or the members may coalesce into a successful fighting unit, daring, stout-hearted, determined. The thoughts and emotions of the individuals might be different, even the outward personalities. But beneath these the members think and react, especially in combat, as a single organism. "Regiments more than any other units," U.S.Gen. Mark W. Clark, World War II commander and President Emeritus of The Citadel, found, "seem to have personalities." He wrote of the mid-20th century after divisions came to dominate the U.S. Army, but stated that James Fry, author of

Combat Soldier, "became so much a part of this personality (of his regiment) that he could not detach himself psychologically from that regiment even when promoted to assistant division commander."[92]

Successful early 19th century French Marshal Thomas R. Bugeaud felt men become soldiers only after they come to accept the regiment as their home. "A man is not a soldier," he wrote, "until he considers his regiment's colors as he would his village steeple, until he loves his colors, and is ready to put hand to sword every time the honor of the regiment is attacked, until he has confidence in his leaders, in his comrades to right and left." This esprit de corps, Bill Mauldin felt, was "the thing that holds armies together." For Forrest's Confederates, Sacramento was a great leap forward toward achieving that goal of creating their own personality.[93]

J.C. Blanton, later captain

photo courtesy Dover Publications
The Federals were probably not as well dressed as this Union cavlaryman, shown here in 1862.

and commander of Company C, 3d Tennessee (Forrest's Old) Regiment, later wrote that the Battle of Sacramento had "a splendid effect" on their unit, "causing men and officers to confide in and respect each other." Further, they "were convinced that evening that Forrest and Kelley were wise selections for our leaders." The leaders "well sustained the reputation made on the field of Sacramento," Blanton continued. The unit now began to take on the character of its leaders. From this point on, as Harry W. Rhodes, of the Confederate Bluff City Grays (afterward Company D, 26th Tennessee Cavalry Battalion in General Forrest's division), pointed out in 1895: "So great was the almost resistless force of his (Forrest's) individual magnetism that he impressed every man in his command with the firm conviction that victory would perch upon his standard ere the battle was fought, and no leader was ever followed to battle with blinder confidence on the part of his soldiery."[94]

Nathan Bedford Forrest, on the other hand, received a demonstration of his men's readiness and ability to fight. Persian King Xerxes is said to have found, at Thermopylae, that he had plenty of combatants—but very few warriors. At Sacramento Forrest found that not only did he have combatants he also had warriors.[95]

On the other side, for the Federals involved Sacramento was not a success. Unfortunately, we do not have post-battle memoirs from the Union participants that describe their feelings. What is known, however, points to many of them having acquitted themselves well, having at times fought with desperate courage. And they could console themselves with the knowledge that they had been not only surprised but outnumbered, although they did overestimate their attackers' size. Several officers, including their young commander, had shown leadership in the face of the

artwork from *Battles and Leaders*, Vol. 3
Col. James Jackson commanded the 3d Kentucky Cavalry in December 1861.

surprise attack. (The sudden attack emphasizes a remark by Maj. Gen. Sir Robert Laycock, Chief of British Combined Operations [Commandos] at the end of World War II "...blessed is he that has his quarrel just, but thrice armed he who gets his blow in first.") Like their Southern opponents, the Union troopers were learning to wage war; and, more importantly, learning to learn the ways of war. For instance, it is highly unlikely the Union survivors would ever again be caught so unprepared while watering their horses.[96]

> "...I have found again and again that in encounter actions, the day goes to the side that is the first to plaster its opponent with fire. The man who lies low and awaits developments usually comes off second best... It is fundamentally wrong simply to halt and look for cover without opening fire, or to wait for more forces to come up and take part in the action."
>
> German Field Marshal Erwin Rommel ("The Desert Fox"), Erwin Rommel, *The Rommel Papers* B.H. Liddell-Hart, ed., 1953, paper rep. N.Y., n.d., p. 7

To the Confederates in Hopkinsville, the Southern victory was one of great joy, particularly as the Federals were part of "Jim Jackson's cavalry," the 3d Kentucky. "There is great rejoicing among southern rights men this evening," a Southerner, writing from Hopkinsville on December 29, exalted, "partly because Jackson is from this place and has often made his boast that he would dine here with his friends on Christmas day." A writer with the 3d Mississippi Infantry, also based at Hopkinsville, reported the Mississippians jealous at Forrest's victory since they too had an "eagerness to measure blades with Col. Jackson, who swore in his wrath that he would dine in Hopkinsville on Christmas Day."[97]

But the victory also propelled Bedford Forrest into notice as a commander. Gen. Albert Sidney Johnston, commanding Confederate forces west of the Appalachians, wrote in his report of operations in December 1861; "For the skill, energy, and courage displayed by Colonel Forrest [at Sacramento] he is entitled to the highest praise, and I take great pleasure in calling the attention of the government to his services."[98]

John Headley, who later served as a scout with Confederate Brig. Gen. John Hunt Morgan's command and later still operated undercover in the North, summed up the Sacramento skirmish when he wrote: "Though an insignificant affair it was regarded at the time as the most sensational and romantic fight of the war, and the daring and intrepidity of Forrest in this, his first battle, brought him instant fame." Basil W. Duke, John Hunt Morgan's justly famous deputy, commented that he first saw Forrest "just after his name had become widely known on account of the victory at

Sacramento."[99]

In the case of both Forrest, who immediately attacked and ended up besting his opponents at Sacramento, and Murray, who managed, although outnumbered, to delay the attack and in the end bring off his troops, leadership shown through. And leadership, as historian and veteran Lt. Col. Richard

> "War is intrinsically harsh and cruel, and blood and tears are its companions..."
>
> Israeli Maj. Gen. Yitshak Rabin, quoted in Libbie L. Braverman and Samuel M. Silver, *The Six-Day Warriors*, N.Y., 1969, p. 59

Hooker wrote, "not machines, not doctrine, not even logistics—is the supreme element in war. Successful commanders throughout history won because they imposed their will on their units, on the enemy, on the battle itself."[100]

As to the course of the war, the skirmish at Sacramento could be said to have had little major significance. Less than 500 men were involved, no territory changed hands, and the course of the war was not appreciably changed. But to a person killed or permanently injured in combat, whether that combat took place in a small skirmish or a large battle is irrelevant. The results for the soldier is the same. As Napolean said, the destruction of an army is a shame, the death of a single soldier is a tragedy. And the battle at Sacramento did see the commencement of outstanding military careers for three future generals and six future colonels. Confederate Bedford Forrest finished the war as a lieutenant general; Confederate Adam Johnson and Federal Eli Murray as brigadier generals; Confederates James W. Starnes, D.C. Kelley, W.S. McLemore, Robert M. Martin and Nicholas C. Gould and Union Robert King as colonels. [101]

One trooper did see the fight in a light many others, perhaps, would soon appreciate. "I was now to realize, in my first actual experience," Sgt. Thomas D. Duncan, of Forrest's command, later wrote, "the fullness of the horrors that wait upon the tinsel glory of that long-worshipped art of human destruction which men call 'war'." Sergeant Duncan, found, as others have found, that "Wounds and death are the currency of war."[102]

- end -

Footnotes

Author's Note: These footnotes tend to go against conventional usage in that "Ibid." is almost nonexistent. In many works it sometimes becomes nearly impossible to find the initial citing for an "Ibid." At the very least it becomes time consuming trying to find that first citation. Here, therefore, the entire citation is used each time so that those who wish to find where a fact came from do not have to hunt forever to find the full citation.

1. For a listing and description of known wars from c. 2000 B.C. to 1986 A.D., see George C. Kohn, *Dictionary of Wars*, rev. ed., N.Y., 1999; John Williams, *World Atlas of Weapons & War*, London, Eng., 1976.

2. Thomas Jordan and J.P. Pryor, *The Campaigns of Lieut.-Gen. N.B. Forrest, and of Forrest's Cavalry*, 1868, rep. Dayton, Ohio, 1973, pp. 44, 49; William D. McCain, "Nathan Bedford Forrest: An Evaluation," *Journal of Mississippi History*, Vol. 24, No. 4 (Oct. 1962), pp. 203-225, pp. 203-204; Otto A. Rothert, *A History of Muhlenberg County*, Louisville, 1913, p. 258; Ed Porter Thompson, *History of the Orphan Brigade*, (1898, rep. Dayton, Ohio, 1973), p. 881.

3. Mark Grimsley, "Millionaire Rebel Raider: The Life of Nathan Bedford Forrest," part 1, *Civil War Times Illustrated*, Oct. 1993, pp. 58-73, pp. 58-64; Robert Selph Henry, *"First With the Most" Forrest*, 1944, rep. N.Y., 1991, pp. 13-34; Thomas Jordan and J.P. Pryor, *The Campaigns of Lieut.-Gen. N.B. Forrest, and of Forrest's Cavalry*, 1868, rep. Dayton, Ohio, 1973, pp. 17-43; William D. McCain, "Nathan Bedford Forrest: An Evaluation," *Journal of Mississippi history*, Vol. 24, No. 4 (Oct. 1962), pp. 203-225, pp. 203-204; W.N.M., "Sketch of Lieutenant-General N.B. Forrest," *Southern Bivouac*, Vol. 2, No. 7 (Mar. 1884), pp. 289-298., pp. 289-292; Ezra J. Warner, *Generals In Gray: Lives of the Confederate Commanders*, Baton Rouge, La., 1959, p. 92; John Allan Wyeth, *That Devil Forrest*, 1899, rep. Baton Rouge, La., 1989, pp. 1-23.

4. John Berrien Lindsley, ed., *The Military Annals of Tennessee, Confederate*, (Nashville, 1886), pp. 761-762; Stewart Sifakis, *Compendium of the Confederate Armies: Tennessee*, N.Y., 1992, p. 80; *Tennesseans In The Civil War*, Nashville, Tenn., 1964, pp. 55-56.

5. Bryan S. Bush, *Lloyd Tilghman: Confederate General in the Western Theatre*, Morley, Mo., 2006, p. 82; Forrest's correspon-

dence, *War of the Rebellion: A Compilation of the Official Records of the Union and Confederate Armies* (commonly *Official Records of the Union and Confederate Armies*), 128 vols., index, and atlases, Washington, D.C., 1880-1901, Ser. I, Vol. 4, p. 551; Robert Selph Henry, *"First With the Most" Forrest,* 1944, rep. N.Y., 1991, pp. 40-42; Thomas Jordan and J.P. Pryor, *The Campaigns of Lieut.-Gen. N.B. Forrest, and of Forrest's Cavalry,* 1868, rep. Dayton, Ohio, 1973, p. 44; Andrew Nelson Lytle, *Bedford Forrest And His Critter Company,* N.Y., 1931, pp. 39-40; John Allan Wyeth, *That Devil Forrest,* 1899, rep. Baton Rouge, La., 1989, pp. 23-25; Marcus Wright, "Memorandum of General Officers," section of "Memorandum of Armies, Corps, and Geographical Command in the Confederate States. 1861-1865," (1876?) in William Frayne Armann, ed., *Personnel of the Civil War,* N.Y., 1961, pp. 235, 357.

6. J.H. Battle, ed.., "Todd County History," J.H. Battle and W.H. Perrin, eds., *Counties of Todd and Christian, Kentucky,* Chicago and Louisville, 1884, p. 110; Forrest's report, *The War of the Rebellion: A Compilation of the Official Records of the Union and Confederate Armies,* 129 vols. and Index, Washington, D.C., 1880-1901, Series I, Vol. 7, pp. 64-65; Louisa H.A. Minor, *The Meriwethers and Their Connections,* Albany, N.Y., 1892, p. 36; John Allan Wyeth, *That Devil Forrest,* 1899, rep. Baton Rouge, La., 1989, p. 28.

7. J.H. Battle, "Todd County History," J.H. Battle and W.H. Perrin, eds., *Counties of Todd and Christian, Kentucky,* Chicago and Louisville, 1884, p. 110; "From Col. Hawkins's Regiment," *Louisville (Ky.) Daily Journal,* Feb. 10, 1862; John W. Headley, *Confederate Operations in Canada and New York,* N.Y., 1906, p. 30; Kentucky Adjutant General, *Report of the Adjutant General of the State of Kentucky, Confederate,* Vol. II, Frankfort, Ky., 1918, pp. 64-65, 272-273; Ed Porter Thompson, *History of the Orphan Brigade,* 1898, rep. Dayton, O., 1973, pp. 881, 1009; Tilghman's correspondence, Forrest's correspondence, *War of the Rebellion: A Compilation of the Official Records of the Union and Confederate Armies* (commonly *Official Records of the Union and Confederate Armies*), 128 vols., index, and atlases, Washington, D.C., 1880-1901, Ser. I, Vol. 4, pp. 485-486; Josephine M. Turner, *The Courageous Caroline Founder of the UDC,* Montgomery, Ala., 1965, pp. 27, 59.

8. H. Gerald Starnes, *Forrest's Forgotten Horse Brigadier,* Bowie,

Md., 1995, pp. 1, 3; *Tennesseans In The Civil War,* 2 parts, Nashville, Tenn., 1964, part 1, p. 62; footnote, Pvt. William H. Terry Military File, National Archives.

9. Forrest's report, *War of the Rebellion: A Compilation of the Official Records of the Union and Confederate Armies* (commonly *Official Records of the Union and Confederate Armies*), 128 vols., index, and atlases, Washington, D.C., 1880-1901, Ser. I, Vol. 7, pp. 64-65; H.T. Gray, "Forrest's First Cavalry Fight," *Confederate Veteran,* Vol. 15 (1907), p. 139.

10. H.T. Gray, "Forrest's First Cavalry Fight," *Confederate Veteran,* Vol. 15 (1907), p. 139; H. Gerald Starnes, *Forrest's Forgotten Horse Brigadier,* Bowie, Md., 1995, p. 1.

11. Forrest's report, *War of the Rebellion: A Compilation of the Official Records of the Union and Confederate Armies* (commonly *Official Records of the Union and Confederate Armies*), 128 vols., index, and atlases, Washington, D.C., 1880-1901, Ser. I, Vol. 7, pp. 64-65; H.T. Gray, "Forrest's First Cavalry Fight," *Confederate Veteran,* Vol. 15 (1907), p. 139.

12. "Return of the 5th Division, Department of the Ohio, for part of the month of Jany, 1862," from Crittenden Military Files, National Archives, General Services Administration, Washington, D.C.; Webb Garrison, *A Treasury of Civil War Tales,* Nashville, Tenn., 1988, Chapter 8: "George and Thomas Crittenden Mirrored the Sundered Nation," pp. 41-44; William D. Gilliam Jr., "Family, Friends and Foe," *The Civil War in Kentucky,* supplement to *The* (Louisville, Ky.) *Courier-Journal,* November 1960, pp. 38, 41, 44; Henry E. Simmons, comp., *A Concise Encyclopedia of the Civil War,* (New York, 1965), p. 64

13. "Statement of the Military Services of Thomas L. Crittenden," compiled by the Adjutant General's Office, 28 February 1882, Crittenden Military Files, National Archives.

14. Adam R. Johnson, *The Partisan Rangers of the Confederate States Army,* William J. Davis, ed., 1904, rep. Evansville, Ind., 1971, p. 40 (states they found Union cavalry crossing the bridge but all other sources state the Federal cavalry had gone out the night before); Otto A. Rothert, *A History of Muhlenberg County,* Louisville, Ky., 1913, p. 259; William E. McLean, in his history of the 43d Indiana Infantry, also mentions the pontoon bridge at Calhoun, see William E. McLean, *The 43d Regiment of Indiana Volunteers,* 1903, rep. Salem, Mass., 1998, p. 77.

15. Adam R. Johnson, *The Partisan Rangers of the Confederate*

States Army, William J. Davis, ed., 1904, rep. Evansville, Ind., 1971, p. 40; Otto A.Rothert, *A History of Muhlenberg County*, Louisville, Ky., 1913, pp. 318-319.

16. Hubert Howe Bancroft, *The Works of Hubert Howe Bancroft*, Vol. 16: *History of the North Mexican States and Texas:* Vol. II 1801-1889, San Francisco, 1889, p. 569, note 34; Adam R. Johnson, *The Partisan Rangers of the Confederate States Army*, William J. Davis, ed., 1904, rep. Evansville, Ind., 1971, pp. 1-40; Otto A.Rothert, *A History of Muhlenberg County*, Louisville, Ky., 1913, pp. 318-319; Ezra J. Warner, *Generals In Gray: Lives of the Confederate Commanders*, Baton Rouge, La., 1959, p. 156.

17. Otto A.Rothert, *A History of Muhlenberg County*, Louisville, Ky., 1913, p. 258.

18. Charles W. Button, "Early Engagements With Forrest," *Confederate Veteran*, Vol. 5 (1897), pp. 478-480, p. 479; Robert Selph Henry, *"First With the Most" Forrest*, 1944, rep. N.Y., 1991, p. 44; Thomas Jordan and J.P. Pryor, *The Campaigns of Lieut.-Gen. N.B. Forrest, and of Forrest's Cavalry*, 1868, rep. Dayton, Ohio, 1973, pp. 50-51; W.M. Protheroe, "It Was A Full Moon When... A Researcher's Guide to Civil War Moon Phases," *Blue & Gray Magazine*, July 1987, pp. 35-37, p. 36; Mauldin quote from Bill Mauldin, *Up Front*, Cleveland, Oh., 1946, p. 35; C.E. Wood, *Mud: A Military History*, Washington, D.C., 2006, p. 11.

19. Forrest's report, *War of the Rebellion: A Compilation of the Official Records of the Union and Confederate Armies* (commonly *Official Records of the Union and Confederate Armies*), 128 vols., index, and atlases, Washington, D.C., 1880-1901, Ser. I, Vol. 7, p. 65.

20. William Henry Perrin, ed., *Counties of Christian and Trigg*, Chicago & Louisville, 1884, p. 188; Union Soldiers and Sailors Monument Association, *The Union Regiments of Kentucky*, Louisville, 1897, pp. 63, 133-134.

21. *Biographical Encyclopaedia of Kentucky of the Dead and Living Men of the Nineteenth Century*, Cincinnati, 1878; H.H. Crittenden, comp., *The Crittenden Memoirs*, N.Y., 1936, pp. 358-359; Thomas Marshall Green, *Historic Families of Kentucky*, Cincinnati, 1889, rep. Baltimore, Md., 1964, pp. 250-251; James S. Jackson Military and Eli H. Murray Military and Pension Files, National Archives; Kentucky Adjutant General, *Report of the Adjutant General of Kentucky, 1861-1866*, 2 vols., Frankfort, Ky., 1866-1867, Vol. 1, p. 80; William Henry Perrin, ed., *Counties*

of Christian and Trigg, Chicago & Louisville, 1884, p. 188; quote, Union Soldiers and Sailors Monument Association, *The Union Regiments of Kentucky,* Louisville, 1897, pp. 63, 133-134.

22. Aloma Williams Dew, " 'Between the Hawk and the Buzzard': Owensboro During the Civil War," *The Register of the Kentucky Historical Society,* Vol. 77, No. 1 (Winter 1979), pp. 1-14, p. 3; Frederich H. Dyer, *A Compendium of the War of the Rebellion,* Vol. III, Regimental Histories, 1908, rep. N.Y., 1959, p. 1191; James S. Jackson, Eli H. Murray Personnel Files, National Archives; Kentucky Adjutant General, *Report of the Adjutant General of Kentucky, 1861-1866,* 2 vols., Frankfort, Ky., 1866-1867, Vol. 1, p. 80; William Henry Perrin, ed., *Counties of Christian and Trigg,* Chicago & Louisville, 1884, p. 188; Union Soldiers and Sailors Monument Association, *The Union Regiments of Kentucky,* Louisville, 1897, pp. 63, 133-134.

23. James Maynard Shanklin, *"Dearest Lizzie": Civil War Letters of James Maynard Shanklin,* Kenneth P. McCutchan, ed., Evansville, Ind., 1988, p. 44.

24. "An Account of the Battle of Sacramento, Kentucky," *Daily Louisville* (Ky.) *Democrat,* Jan. 7, 1862, reprinted from *Colombia* (Mo.) *Statesman,* Dec. 30, 1862; "An Affair At Sacramento," *Louisville* (Ky.) *Daily Journal,* Jan. 7, 1862; "Cavalry Skirmish," *Evansville* (Ind.) *Journal,* Dec. 30, 1861; Crittenden's and Forrest's reports, *War of the Rebellion: A Compilation of the Official Records of the Union and Confederate Armies* (commonly *Official Records of the Union and Confederate Armies*), 128 vols., index, and atlases, Washington, D.C., 1880-1901, Ser. I, Vol. 7, pp. 62-66; exact companies ascertained by casualties' units of assignment; Larry J. Daniel, *Days of Glory: The Army of the Cumberland 1861-1865,* Baton Rouge, La., 2004, p. 37; Clarence Linden, "Particulars of the Sacramento Skirmish," *Louisville* (Ky.) *Journal,* Feb. 24, 1862; Charles Mayfield Meacham, *A History of Christian County, Kentucky from Oxcart to Airplane,* Nashville, Tenn., 1930, pp. 143-144; Military Affairs In Kentucky," *Western Citizen,* Paris, Ky., Jan. 10, 1862; Sherman's correspondence, *War of the Rebellion: A Compilation of the Official Records of the Union and Confederate Armies* (commonly *Official Records of the Union and Confederate Armies*), 128 vols., index, and atlases, Washington, D.C., 1880-1901, Ser. I, Vol. 7, p. 324; "The Skirmish At Sacramento, Ky.," *Chicago Times,* Jan. 6, 1862; H. Gerald

Starnes, *Forrest's Forgotten Horse Brigadier,* Bowie, Md., 1995, p. 1.

25. Otto A.Rothert, *A History of Muhlenberg County,* Louisville, Ky., 1913, p. 259; "Cavalry Skirmish," *Evansville* (Ind.) *Journal,* Dec. 30, 1861; J. Harold Utley, "The Battle of Sacramento... With Nathan Bedford Forrest & Madisonville's Al Fowler," Battle of Sacramento Planning Committee, *"Down Memory Lane" in Sacramento, Kentucky,* n.p., 1999, unnumbered pages.

26. Forrest's report, *War of the Rebellion: A Compilation of the Official Records of the Union and Confederate Armies* (commonly *Official Records of the Union and Confederate Armies*), 128 vols., index, and atlases, Washington, D.C., 1880-1901, Ser. I, Vol. 7, p. 65; Adam R. Johnson, *The Partisan Rangers of the Confederate States Army,* William J. Davis, ed., 1904, rep. Evansville, Ind., 1971, p. 41; John Allan Wyeth, *That Devil Forrest,* 1899, rep. Baton Rouge, La., 1989, p. 28.

27. Exact identity of this young lady was lost until Elizabeth Cox fully identified her in 1971. Jordan and Pryor gave her name only as Miss Morehead (sic); Johnson describes the incident but has no name; Ms. Cora Bennett, "Correspondence," *McLean Co. News* (Calhoun, Ky.), February 22, 1962, gives the names of some of her relatives; Elizabeth Cox, McLean Co. genealogist, used these facts to first identify the girl and her sister, contained in correspondence to the author, 1971. Forrest's difficulty in keeping the young lady from riding into combat with him is described in Adam R. Johnson, *The Partisan Rangers of the Confederate States Army,* William J. Davis, ed., 1904, rep. Evansville, Ind., 1971, p. 41; Thomas Jordan and J.P. Pryor, *The Campaigns of Lieut.-Gen. N.B. Forrest, and of Forrest's Cavalry,* 1868, rep. Dayton, Ohio, 1973, p. 50; Robert Selph Henry, *"First With the Most" Forrest,* 1944, rep. N.Y., 1991, p. 44; Antonio Fraser, *The Warrior Queens,* 1988, rep. N.Y., 1989, p. 139.

28. Shelley Saywell, *Women In War,* N.Y., 1985, p. ix; Jean V. Berlin, "Introduction to the Bison Book Edition," *Women in the Civil War,* originally *Bonnet Brigades,* 1966, rep. Lincoln, Neb., 1994, pp. vii-xvi, p. vii; also see Deanne Blanton and Lauren M. Cook, *They Fought Like Demons: Women Soldiers in the Civil War,* 2002, rep. N.Y. 2003.

29. Adam R. Johnson, *The Partisan Rangers of the Confederate States Army,* William J. Davis, ed., 1904, rep. Evansville, Ind., 1971, p. 41; Wilena Roberts Bejach, "Civil War Letters of a Mother

and Son," *The West Tennessee Historical Society Papers*, No. 4, 1950. , pp. 50-71, p. 52.

30. H.T. Gray, "Forrest's First Cavalry Fight," *Confederate Veteran*, Vol. 15 (1907), p. 139.

31. James L. Stokesbury, "Introduction," in Martin Blumenson and James L. Stokesbury, *Masters of the Art of Command*, 1975, rep. N.Y., n.d., p. 2; Adam R. Johnson, *The Partisan Rangers of the Confederate States Army*, William J. Davis, ed., 1904, rep. Evansville, Ind., 1971, p. 42; Thomas Jordan and J.P. Pryor, *The Campaigns of Lieut.-Gen. N.B. Forrest, and of Forrest's Cavalry*, 1868, rep. Dayton, Ohio, 1973, p. 51.

32. Adam R. Johnson, *The Partisan Rangers of the Confederate States Army*, William J. Davis, ed., 1904, rep. Evansville, Ind., 1971, p. 42; Lester N. Fitzhugh, "One Texas Unit Served Forrest," clipping from unidentified newspaper; Wilena Roberts Bejach, "Civil War Letters of a Mother and Son," *West Tennessee Historical Society Papers*, 1950, No. 4, p. 52, letter from James H. Hamner, a trooper under Forrest.

33. Ernest Hemingway, "Introduction," Ernest Hemingway, ed., *Men At War*, 1942, rep. N.Y., 1955, p. xix; Admiral Mahan quoted in Theordore Roscoe, *Pig Boats* (originally published as *United States Submarine Operations in World War II*), rep. N.Y., 1958, p. 64; John Keegan and Richard Holmes, *Soldiers: A History of Men In Battle*, N.Y., 1985, p. 208; Forrest's report, *War of the Rebellion: A Compilation of the Official Records of the Union and Confederate Armies* (commonly *Official Records of the Union and Confederate Armies*), 128 vols., index, and atlases, Washington, D.C., 1880-1901, Ser. I, Vol. 7, Ser. I, Vol. 7, p. 65; Crittenden's report, Ibid., p. 62.

34. Forrest's report, *War of the Rebellion: A Compilation of the Official Records of the Union and Confederate Armies* (commonly *Official Records of the Union and Confederate Armies*), 128 vols., index, and atlases, Washington, D.C., 1880-1901, Ser. I, Vol. 7, p. 65; Thomas Jordan and J.P. Pryor, *The Campaigns of Lieut.-Gen. N.B. Forrest, and of Forrest's Cavalry*, 1868, rep. Dayton, Ohio, 1973, p. 51; Alan Axelrod, *Patton On Leadership: Strategic Lessons For Corporate Warfare*, Paramus, N.J., 1999, p. 108.

35. Forrest's report, *War of the Rebellion: A Compilation of the Official Records of the Union and Confederate Armies* (commonly *Official Records of the Union and Confederate*

Armies), 128 vols., index, and atlases, Washington, D.C., 1880-1901, Ser. I, Vol. 7, p. 65, does not give details of Meriwether's death; Thomas Jordan and J.P. Pryor, *The Campaigns of Lieut.-Gen. N.B. Forrest, and of Forrest's Cavalry,* 1868, rep. Dayton, Ohio, 1973, p. 51; Louisa H.A. Minor, *The Meriwethers and Their Connections,* Albany, N.Y., 1892, p. 132.

36. J.R. Chalmers, "Forrest and his Campaigns," *Southern Historical Society Papers,* Vol. 7 (1879), pp. 451-486, p. 456.

   Leuthen and Zorndorf were classic battles studied by anyone interested in the military prior to and immediately following the Civil War. Today they are virtually forgotten. Leuthen occurred on December 5, 1757, during the Seven Years' War and is considered Prussian Frederick the Great's greatest victory. With 36,000 men Frederick attacked 80,000 entrenched Austrians under Marshal Prince Charles of Lorraine and Marshal Count Leopold von Daun. The Austrians scattered in retreat after losing 20,000 captured, 6,750 killed or wounded and 116 guns and 51 colors taken; the Prussians lost 6,150 killed or wounded. The Battle of Leuthen "Showed (Frederick's) complete tactical mastery in the field..." Zorndorf (in present Poland) was fought on August 25, 1758, also during the Seven Years' War. Frederick led a 36,000-man Prussian-Hanovian army against a 40,000-man Russian army under Count William Fermor threatening Prussia. Frederick's army was saved from almost certain disaster by Prussian cavalry Gen. Friedrich Wilhelm von Seydlitz's cavalry suddenly falling on the Russian's right flank and rear. General Seydlitz has been described as "a charismatic leader and brilliant battlefield tactician with an uncanny sense of timing." See Michael Calvert, with Peter Young, *A Dictionary of Battles: 1715-1815,* N.Y., 1979, pp. 26, 45; also, for Zorndorf and General Seydlitz, Eric Niderost, "Mauled By The Russian Bear," *Military Heritage,* April 2006, pp. 26-33, 67.

37. Sun Tzu, *The Art of War,* trans. with intro. by Marine Brig. Gen. Samuel B. Griffith (ret.), N.Y., 1963, p. 135 and note.

38. Pearl J. Riser, "Nathan Bedford Forrest: 'The Wizard of the Saddle'," *United Daughters of the Confederacy Magazine,* April 1950, pp. 11-14, p. 14.

39. J.R. Chalmers, "Forrest and his Campaigns," *Southern Historical Society Papers,* Vol. 7 (1879), pp. 451-486, pp. 456-457; axiom in, among other sources, Martin Blumenson and James L. Stokesbury, *Masters of the Art of Command,* 1975, rep. N.Y., n.d., p. 221.

40. Crittenden's report, *War of the Rebellion: A Compilation of*

the Official Records of the Union and Confederate Armies (commonly *Official Records of the Union and Confederate Armies*), 128 vols., index, and atlases, Washington, D.C., 1880-1901, Ser. I, Vol. 7, pp. 62-63; Pierre Leulliette, *The War in Algeria: Memoirs of a Paratrooper,* 1961, paperback rep. N.Y., 1987, p. 200.

41. "Captain A.N. Davis...," *Daily Commonwealth,* Aug. 30, 1862.

42. Carl von Clausewitz, *On War,* Michael Howard and Peter Paret, eds. and trans., Princeton, N.Y., 1976, reprinted with new essays, 1984, p. 76; Samuel B. Griffith, "Introduction," Sun Tzu, *The Art of War,* trans. with intro. by Marine Brig. Gen. Samuel B. Griffith (ret.), N.Y., 1963., p. v; John de St. Jorre, "The Imperial War Museum," *MHQ: The Quarterly Journal of Military History,* Vol. 3, No. 3 (Spring 1991), pp. 8-16, p. 11; Charles W. Button, "Early Engagements With Forrest," *Confederate Veteran,* Vol. 5 (1897), pp. 478-480, p. 479; John Steinbeck, *Once There Was A War,* 1958, rep. N.Y., 1960, p. 8.

It would, in fact, have been profoundly improbable for Forrest to have heard of von Clausewitz. Until 1871 the writer was known only in Prussia, where his book had been published in the 1830s. Even after 1871 and until World War I translations are considered flawed and incomplete. See Larry H. Addington, *The Patterns of War Since The Eighteenth Century,* Bloomington, Ind., 1984, pp. 41-44, and Martin Van Creveld, *The Art of War: War and Military Thought,* 2000, rep. N.Y., 2005, pp. 107-115.

43. (Plutarch?), "The Sayings of Spartans," in Plutarch, *Plutarch On Sparta,* trans. with intro and notes by Richard J.A. Talbert, N.Y., 1988, pp. 109-157, p. 124.

44. Forrest's report, *War of the Rebellion: A Compilation of the Official Records of the Union and Confederate Armies* (commonly *Official Records of the Union and Confederate Armies*), 128 vols., index, and atlases, Washington, D.C., 1880-1901, Ser. I, Vol. 7, p. 65; Thomas Jordan and J.P. Pryor, *The Campaigns of Lieut.-Gen. N.B. Forrest, and of Forrest's Cavalry,* 1868, rep. Dayton, Ohio, 1973, p. 52, and Adam R. Johnson, *The Partisan Rangers of the Confederate States Army,* William J. Davis, ed., 1904, rep. Evansville, Ind., 1971, p. 42, describe the melee; Waugh diary quoted in Paul Fussell, *Wartime: Understanding and Behavior in the Second World War,* 1989, rep. N.Y., 1990, p. 4.

45. "Capt. Bacon," *Evansville* (Ind.) *Journal,* Jan. 3, 1862; "Death of

Capt. Bacon," *Louisville* (Ky.) *Journal*, Jan. 1, 1862; "Particulars of the Sacramento Skirmish," *Louisville* (Ky.) *Daily Journal*, Feb. 24, 1862; Otto A.Rothert, *A History of Muhlenberg County*, Louisville, Ky., 1913, p. 296; "The Skirmish at Sacramento, Kentucky," *Chicago* (Ill.) *Times*, Jan. 6, 1862, citing the *Louisville* (Ky.) *Democrat*.

46. Charles W. Button, "Early Engagements With Forrest," *Confederate Veteran*, Vol. 5 (1897), pp. 478-480, p. 479; Forrest's report, *War of the Rebellion: A Compilation of the Official Records of the Union and Confederate Armies* (commonly *Official Records of the Union and Confederate Armies*), 128 vols., index, and atlases, Washington, D.C., 1880-1901, Ser. I, Vol. 7, p. 65; Gellhorn, Martha, *The Face Of War*, rev. ed., N.Y., 1988, p. 146; Martha Gellhorn, "The Battle of the Bulge," in Jon E. Lewis, ed., *The Mammoth Book of True War Stories*, N.Y., 1992, pp. 353-360, p. 354; Napoleon I, *The Military Maxims of Napoleon*, Lt. Gen. Sir George C. D'Aguilar, David G. Chandler, commentary, 1831, rep. Mechanicsburg, Pa., 2002, p. 73.

47. Kentucky Adjutant General, *Report of the Adjutant General of Kentucky, 1861-1866*, Vol. 1, Frankfort, Ky., 1866-1867, p. 64; *Medical and Surgical History of the War of the Rebellion*, Joseph K. Barnes, ed., 6 vols., Washington, D.C., 1870-1888, Vol. 2, pt. 2, p. 76 (listed as Dec. 1862).

48. *History of Daviess County, Kentucky, 1883*, rep. Evansville, Ind., 1966, p. 622; Kentucky Adjutant General, *Report of the Adjutant General of Kentucky, 1861-1866*, 2 vols., Frankfort, Ky., 1866-1867, Vol. 1, p. 67; Barbara Sistler; Byron Sistler and Samuel Sistler, *1850 Census, North West Kentucky*, Nashville, Tenn. 1994, p. 52.

49. Thomas Jordan and J.P. Pryor, *The Campaigns of Lieut.-Gen. N.B. Forrest, and of Forrest's Cavalry*, 1868, rep. Dayton, Ohio, 1973, footnote, pp. 53-54; Martin Blumenson and James L. Stokesbury, *Masters of the Art of Command*, 1975, rep. N.Y., n.d., p. 41.

50. Clark's report, *War of the Rebellion: A Compilation of the Official Records of the Union and Confederate Armies* (commonly *Official Records of the Union and Confederate Armies*), 128 vols., index, and atlases, Washington, D.C., 1880-1901, Ser. I, Vol. 7, p. 64.

51. Roger Ford and Tim Ripley, *The Whites of Their Eyes: Close-Quarter Combat*, 1997, rep. Dulles, Va., 2001, "Introduction," pp.

4-5.

52.   Adam R. Johnson, *The Partisan Rangers of the Confederate States Army,* William J. Davis, ed., 1904, rep. Evansville, Ind., 1971, p. 142; Mitchell Military File, National Archives.

53.   Richard T. Martin, "Recollections of the Civil War," in Otto A. Rothert, *A History of Muhlenberg County,* Louisville, 1913, pp. 285-317, p. 297; Otto A. Rothert, *A History of Muhlenberg County,* Louisville, 1913, p. 300.

54.   Ernle Bradford, *Thermopylae: The Battle For the West,* 1980, rep. Cambridge, Mass., n.d., p. 109.

55.   Michael Cotten, *The Williamson County Cavalry: A History of Company F, Fourth Tennessee Cavalry Regiment, C.S.A.,* n.p., 1994, pp. 15-16, 33-34; Thomas Jordan and J.P. Pryor, *The Campaigns of Lieut.-Gen. N.B. Forrest, and of Forrest's Cavalry,* 1868, rep. Dayton, Ohio, 1973, p. 54; Forrest's report, *War of the Rebellion: A Compilation of the Official Records of the Union and Confederate Armies* (commonly *Official Records of the Union and Confederate Armies*), 128 vols., index, and atlases, Washington, D.C., 1880-1901, Ser. I, Vol. 7, p. 66; Terry Military File, National Archives; D.W. Johnston, "Correspondence," *Evansville* (Ind.) *Journal,* Jan. 28, 1862; B.L. Ridley, "Chat With Col. W.S. McLemore," *Confederate Veteran,* Vol. 8 (1900), p. 262; H. Gerald Starnes, *Forrest's Forgotten Horse Brigadier,* Bowie, Md., 1995, p. 2.

56.   Crittenden's report, *War of the Rebellion: A Compilation of the Official Records of the Union and Confederate Armies* (commonly *Official Records of the Union and Confederate Armies*), 128 vols., index, and atlases, Washington, D.C., 1880-1901, Ser. I, Vol. 7, p. 63; Forrest's report, Ibid., pp. 65-66; Davis Invalid Pension Application, Pension File, National Archives; Richard T. Martin, "Recollections of the Civil War," in Otto A. Rothert, *A History of Muhlenberg County,* Louisville, 1913, pp. 285-317, pp. 296-297, pp. 296-297.

57.   Thomas Jordan and J.P. Pryor, *The Campaigns of Lieut.-Gen. N.B. Forrest, and of Forrest's Cavalry,* 1868, rep. Dayton, Ohio, 1973, p. 53; *Spirit of the Age* (Raleigh, N.C.), Jan. 15, citing the *Nashville* (Tenn.) *Banner,* states that Col. Starnes shot Bacon; item reprinted in Frank Moore, ed., *The Rebellion Record: A Diary of American Events, With Documents, Narratives, Illustrative Incidents, Poetry, ETC.,* 11 vols., (New York, 1862), Vol. 3,

p. 517.

58. *Chicago* (Ill.) *Times,* Jan. 6, 1862; Crittenden's report, *War of the Rebellion: A Compilation of the Official Records of the Union and Confederate Armies* (commonly *Official Records of the Union and Confederate Armies*), 128 vols., index, and atlases, Washington, D.C., 1880-1901, Ser. I, Vol. 7, p. 63; Walters' Military File, National Archives.

59. Roger Ford and Tim Ripley, *The Whites of Their Eyes: Close-Quarter Combat,* 1997, U.S.A. rep. Dulles, Va., 2001, p. 107.

60. Adam R. Johnson, *The Partisan Rangers of the Confederate States Army,* William J. Davis, ed., 1904, rep. Evansville, Ind., 1971, p. 43; Thomas Jordan and J.P. Pryor, *The Campaigns of Lieut.-Gen. N.B. Forrest, and of Forrest's Cavalry,* 1868, rep. Dayton, Ohio, 1973, p. 53.

61. Adam R. Johnson, *The Partisan Rangers of the Confederate States Army,* William J. Davis, ed., 1904, rep. Evansville, Ind., 1971, p. 43; Thomas Jordan and J.P. Pryor, *The Campaigns of Lieut.-Gen. N.B. Forrest, and of Forrest's Cavalry,* 1868, rep. Dayton, Ohio, 1973, p. 53; H. Gerald Starnes, *Forrest's Forgotten Horse Brigadier,* Bowie, Md., 1995, p. 2.

62. Adam R. Johnson, *The Partisan Rangers of the Confederate States Army,* William J. Davis, ed., 1904, rep. Evansville, Ind., 1971, p. 43; Thomas Jordan and J.P. Pryor, *The Campaigns of Lieut.-Gen. N.B. Forrest, and of Forrest's Cavalry,* 1868, rep. Dayton, Ohio, 1973, p. 53; Wilena Roberts Bejach, "Civil War Letters of a Mother and Son," *The West Tennessee Historical Society Papers,* No. 4, 1950. , pp. 50-71, p. 52; H.T. Gray, "Forrest's First Cavalry Fight," *Confederate Veteran,* Vol. 15 (1907), p. 139.

63. Adam R. Johnson, *The Partisan Rangers of the Confederate States Army,* William J. Davis, ed., 1904, rep. Evansville, Ind., 1971, p. 43.

64. "From Kentucky," *Memphis* (Tenn.) *Daily Appeal,* Jan. 8, 1862; "From Kentucky," *Memphis* (Tenn.) *Daily Appeal,* Jan. 9, 1862.

65. Individual Military Files, National Archives; Kentucky Adjutant General, *Report of the Adjutant General of Kentucky 1861-1866,* 2 vol. (Frankfort, Ky.), Vol. 1, pp. 63-69; Otto A.Rothert, *A History of Muhlenberg County,* Louisville, Ky., 1913, pp. 296-297; Crittenden's report, *War of the Rebellion: A Compilation of the Official Records of the Union and Confederate Armies* (commonly *Official Records of the Union and Con-*

federate Armies), 128 vols., index, and atlases, Washington, D.C., 1880-1901, Ser. I, Vol. 7, p. 63; "Local Laconics," *Owensboro* (Ky.) *Daily Messenger,* Oct. 7, 1894.

66. Jayne E. Blair, *The Essential Civil War: A Handbook to the Battles, Armies, Navies and Commanders,* Jefferson, N.C., 2006, p. 74; Crittenden Report, *War of the Rebellion: A Compilation of the Official Records of the Union and Confederate Armies* (commonly *Official Records of the Union and Confederate Armies*), 128 vols., index, and atlases, Washington, D.C., 1880-1901, Ser. I, Vol. 7, p. 63; Captain A.N. Davis to President A. Lincoln, Correspondence dated Salisbury, N. C., April 2, 1862, Ibid., Ser. II, Vol. 3, pp. 418-419, requesting exchange consideration; "Fight At Sacramento, Ky.," *Chicago* (Ill.) *Times,* January 3, 1862.

67. R.T. Martin, "Recollections of the Civil War," Otto A.Rothert, *A History of Muhlenberg County,* Louisville, Ky., 1913, pp. 285-317, p. 297.

68. Baker Military File, National Archives.

69. "From Kentucky," *Memphis* (Tenn.) *Daily Appeal,* Jan. 9, 1862.

70. John McElroy, *Andersonville: A Story of Rebel Prisons,* 2 vols. , 1879, rep. Washington, D.C., 1899, Vol. 1, p. 60; Rev. T.J. Sheppard, "Religious Life and Work in Andersonville—How Captured...," in John McElroy, *Andersonville: A Story of Rebel Prisons,* 2 vols. , 1879, rep. Washington, D.C., 1899, Vol. 2, pp. 628-638, p. 628.

71. Forrest's report, *War of the Rebellion: A Compilation of the Official Records of the Union and Confederate Armies* (commonly *Official Records of the Union and Confederate Armies*), 128 vols., index, and atlases, Washington, D.C., 1880-1901, Ser. I, Vol. 7, pp. 64-65; Clark's report, Ibid., p. 64.

72. Kentucky Adjutant General, *Report of the Adjutant General of Kentucky, 1861-1866,* 2 vols., Frankfort, Ky., 1866-1867, Vol. 1, pp. 63-80, 441-432; *War of the Rebellion: A Compilation of the Official Records of the Union and Confederate Armies* (commonly *Official Records of the Union and Confederate Armies*), 128 vols., index, and atlases, Washington, D.C., 1880-1901, Ser. I, Vol. 7, pp. 62-66.

73. Crittenden's report, *War of the Rebellion: A Compilation of the Official Records of the Union and Confederate Armies* (commonly *Official Records of the Union and Confederate Armies*), 128 vols., index, and atlases, Washington, D.C., 1880-

1901, Ser. I, Vol. 7, p. 63.

74. Ernle Bradford, *Thermopylae: The Battle For the West,* 1980, rep. Cambridge, Mass., n.d., p. 35.

75. *McLean County, Kentucky 1860 Census Annotated,* Owensboro, Ky., 1978, Holly M. Leftwich, transcriber, Elizabeth S. Cox, annotater; "More Prisoners," *Louisville* (Ky.) *Daily Journal,* January 11, 1862; "The Sacramento Affair," *Louisville* (Ky.) *Democrat,* Jan. 7, 1862; H. Gerald Starnes, *Forrest's Forgotten Horse Brigadier,* Bowie, Md., 1995, p. 2; *War of the Rebellion: A Compilation of the Official Records of the Union and Confederate Armies* (commonly *Official Records of the Union and Confederate Armies*), 128 vols., index, and atlases, Washington, D.C., 1880-1901, Ser. I, Vol. 7, p. 66.

76. Benson J. Lossing, *Mathew Brady's Illustrated History of the Civil War* (originally, *A History of the Civil War*), 1912, rep. N.Y., n.d., p. 64; Rossiter Johnson,*Campfire and Battlefield,*1894, rep. N.Y., 1978, p. 115.

77. Wilena Roberts Bejach, "Civil War Letters of a Mother and Son," *The West Tennessee Historical Society Papers,* No. 4, 1950. , pp. 50-71, p. 53; Roger Ford and Tim Ripley, *The Whites of Their Eyes: Close-Quarter Combat,* 1997, U.S.A. rep. Dulles, Va., 2001, pp. 249-250.

78. Webb Garrison, *Friendly Fire in the Civil War,* Nashville, Tenn., 1999; also see Robert F. Dorr, "When Americans Killed Americans," in *America In WWII,* Vol. 5, No. 5 (Jan-Feb., 2010), pp. 50-57.

79. Forrest's Report, *War of the Rebellion: A Compilation of the Official Records of the Union and Confederate Armies* (commonly *Official Records of the Union and Confederate Armies*), 128 vols., index, and atlases, Washington, D.C., 1880-1901, Ser. I, Vol. 7, p. 66.

80. "The Affair at Sacramento," *Frankfort* (Ky.) *Yeoman,* Jan. 2, 1862; Forrest's report, *War of the Rebellion: A Compilation of the Official Records of the Union and Confederate Armies* (commonly *Official Records of the Union and Confederate Armies*), 128 vols., index, and atlases, Washington, D.C., 1880-1901, Ser. I, Vol. 7, pp. 64-66; "From Kentucky," *Memphis* (Tenn.) *Daily Appeal,* Jan. 8, 1862; Ann Vance Todd, correspondence with author, Aug. 7, 1994, containing copies of the depositions of Isaac and Elizabeth Johnson; John Steinbeck, *Once There Was A War,* 1958, rep. N.Y., 1960, p. 123.

81. "The Death of Captain Bacon," *Louisville* (Ky.) Daily Journal, Jan. 27, 1862.

82. Robert J. Brandon, *Autobiography of Robert J. Brandon*, with additional notes by Edith Bennett, n.p., [1992], pp. 6-7; Crittenden's report, *War of the Rebellion: A Compilation of the Official Records of the Union and Confederate Armies* (commonly *Official Records of the Union and Confederate Armies*), 128 vols., index, and atlases, Washington, D.C., 1880-1901, Ser. I, Vol. 7., p. 62; Union soldiers and Sailors Monument Association, *The Union Regiments of Kentucky* (Louisville, 1897), p. 539.

83. "The Affair at Sacramento," *Frankfort* (Ky.) *Yeoman*, Jan. 2, 1862; Robert J. Brandon, *Autobiography of Robert J. Brandon*, with additional notes by Edith Bennett, n.p., [1992], p. 7; Crittenden's report, *War of the Rebellion: A Compilation of the Official Records of the Union and Confederate Armies* (commonly *Official Records of the Union and Confederate Armies*), 128 vols., index, and atlases, Washington, D.C., 1880-1901, Ser. I, Vol. 7, pp. 62-63; Thomas Southard, "Sacramento sixth graders report more local Civil War stories," *McLean Co. News*, Calhoun, Ky., Nov. 27, 1958.

84. "Civil War Stories: Station Wounded Union Soldier," Battle of Sacramento Planning Committee, *"Down Memory Lane" in Sacramento, Kentucky*, n.p., 1999, pages unnumbered.

85. Forrest's report, *War of the Rebellion: A Compilation of the Official Records of the Union and Confederate Armies* (commonly *Official Records of the Union and Confederate Armies*), 128 vols., index, and atlases, Washington, D.C., 1880-1901, Ser. I, Vol. 7, , p. 66; "From Kentucky," *Memphis* (Tenn.) *Daily Appeal*, Jan. 8, 1862; William Preston Johnston, *The Life of Gen. Albert Sidney Johnston*, N.Y., 1879, p. 386; Richard T. Martin, "Recollections of the Civil War," in Otto A. Rothert, *History of Muhlenberg County*, Louisville, Ky., 1913, pp. 285-317, p. 296; Otto A.Rothert, *A History of Muhlenberg County*, Louisville, Ky., 1913, p. 262.

86. John W. Muster III, Interview.

87. B.L. Ridley, "Chat With Col. W.S. McLemore," *Confederate Veteran*, Vol. 8 (1900), p. 262; Bromfield L. Ridley, "Chat With Col. W.S. McLemore," *Battles and Sketches of the Army of Tennessee*, Mexico, Mo., 1906, pp. 177-181, p. 178.

88. Gerald Astor with Joseph Millard, "Heller on Horseback," in Joseph Millard, ed., *True Civil War Stories*, Greenwich, Conn., 1961, pp.

169-184, p. 175.

89. T. Harry Williams, "Kentucky, The Hard School of Experience," *The Civil War in Kentucky,* supplement to *The* (Louisville, Ky.) *Courier-Journal,* November 1960, pp. 53-54, 58..

90. Richard Holmes, *Acts of War: The Behavior of Men in Combat,* N.Y., 1985, pp. 74-75; John Hersey, *Of Men And War,* 1944, rep. N.Y., 1963, p. 42; Paul Fussell, *Wartime: Understanding and Behavior in the Second World War,* N.Y., 1989, paper rep., 1990, p. 75; Bill Mauldin, *Up Front,* Cleveland, Oh., 1946, pp. 46-47.

91. Charles R. Anderson, *Vietnam: The Other War,* 1982, rep. N.Y., 1990, p. 218; Peter Maslowski, "Reel War vs. Real War," *MHQ: The Quarterly Journal of Military History,* Vol. 10, No. 4 (Summer 1998), pp. 68-75, p. 74; Paul Dickson, *War Slang,* 2d ed., N.Y., 2000; pp. 9, 19; Webb Garrison, with Cheryl Garrison, *The Encyclopedia of Civil War Usage,* Nashville, Tenn., 2001, under "elephant" and "see the elephant"; John Steinbeck, *Once There Was A War,* 1958, rep. N.Y., 1960, p. 111; John Keegan and Richard Holmes, *Soldiers: A History of Men In Battle,* N.Y., 1985, p. 261; William Breuer, *Devil Boats: The PT War Against Japan,* 1987, rep. N.Y., 1988, p. 112.

92. Mark W. Clark, "Foreword," in James C. Fry, *Combat Soldier,* Washington, D.C., 1968, p. v.

93. Richard Holmes, *Acts of War: The Behavior of Men in Combat,* N.Y., 1985, p. 80; Bill Mauldin, *Up Front,* Cleveland, Oh., 1946, p. 1.

94. J.C. Blanton, "Forrest's Old Regiment," *Confederate Veteran,* Vol. 3 (1895), pp. 41-42, 77-78, p. 41, also quoted in David Knapp Jr., *The Confederate Horsemen,* N.Y., 1966, p. 245; Harry W. Rhodes, "Military Character of Gen. Forrest," *Confederate Veteran,* Vol. 3 (1895), p. 212.

95. Xerxes' findings from Herodotus, "Thermopylae," in John Bettenbender and George Fleming, eds., *Famous Battles,* N.Y., 1970, pp. 43-51. p. 46.

96. Sir Robert Laycock, "Forward," in Charles Foley, *Commando Extraordinary: Otto Skorzeny,* 1954, rep. London, Eng. 1998, p. 8.

97. "From Kentucky," *Memphis* (Tenn.) *Daily Appeal,* Jan. 8, 1862; "From Kentucky," *Memphis* (Tenn.) *Daily Appeal,* Jan. 9, 1862; "From Kentucky," *Memphis* (Tenn.) *Daily Appeal,* Jan. 9, 1862.

98. Curt Anders, *Fighting Confederates,* N.Y., 1968, p. 111.

99. Basil W. Duke, *Reminiscences of General Basil W. Duke,*

C.S.A., N.Y., 1911, p. 345; John W. Headley, *Confederate Operations in Canada and New York,* N.Y., 1906, p. 33.

100. Richard D. Hooker Jr., ed., *By Their Deeds: America's Combat Commanders On The Art Of War,* N.Y., 2003.

101. Napolean quoted in Anthony Arthur, *Bushmasters: America's Jungle Warriors of World War II,* 1987, rep. N.Y. 1989, p. 263.

102. Thomas D. Duncan, *The Recollections of Thomas D. Duncan,* 1922, rep. Savannah, Tenn., 2000, Tony Hays, ed., p. 10; John Keegan and Richard Holmes, *Soldiers: A History of Men In Battle,* N.Y., 1985, p. 141.

Selected Personnel After-Action Activities

Sacramento Area Citizens

The citizens of Sacramento and the skirmish area did not escape the effects of the Battle of Sacramento. Ed Coffman, later a Central City merchant, was a boy living on a farm near Sacramento in December 1861. He had gone to the barn to feed the stock when he heard and saw troops passing. He climbed to the hayloft and hid in the hay. Through the cracks he saw two soldiers "come into the barn and take their finest horse and make off with him." These were probably Confederates, although Ed didn't say.

Some arrests were made of Sacramento citizens for firing at the Federals or otherwise aiding the Confederates at the Battle of Sacramento. One report states the parties were "dealt with." Another report states the citizens were taken to the Federal camp at Calhoun but released upon taking the oath of allegiance to the Union. Still another item lists five Southerners captured at the Battle of Sacramento. These may have been citizens.

On January 16, 1862, General Crittenden moved his division to South Carrollton, upriver in Muhlenberg Co. One brigade, the Thirteenth, went by steamboat and barge in order to swiftly occupy South Carrollton. The Fourteenth Brigade, with the division wagons, and the 3d Kentucky Cavalry moved overland, by way of Sacramento. The 31st Indiana Infantry Regiment moved swiftly forward in order to occupy Sacramento "before the inhabitants were aware of the approach of the troops."

S.F. Horrall, an officer of the 42d Indiana Infantry and historian of the unit, later admitted that the "boys, finding plenty of chickens in the neighborhood, because of this ambush (the skirmish of December 28) believed to have been led by the citizens, raided the hen-roosts far and near, and before guard lines were fixed." The culprits were arrested but General Crittenden's headquarters chaplain "became religiously eloquent for the release of the comrades" and "prevailed."

"The roads were comparatively smooth from Rumsey to Sacramento, where we 'encamped' for the night, taking up quarters in churches,warehouses, school houses, &c., &c.," a member of the 42d, who identified himself as 'Q.K. Juniper Wiggins,' reported. He did not admit any chicken stealing. He did write that the town "denotes thrift—or has been." Now, however, there were only seven families left. These all claimed to be "for the Union." Then, getting Biblical, the soldier pointed out that "it was promised in the days of antiquity, once upon a time, that a certain city should be spared if five righteous could be found therein, so we, finding seven, concluded that Sacramento ought to be 'more than saved'."

Sources: "Capt. Bacon," *Evansville* (Ind.) *Journal*, Jan. 3, 1862; Agnes S. Harralson, *Steamboats on the Green*, Berea, Ky., 1981, p. 154; S.F. Horrall, *The History of the Forty-Second Indiana*, Chicago, Ill., 1892, pp. 105-106; Otto A. Rothert, *A History of Muhlenberg County*, Louisville, Ky., 1913, pp. 263-264; OR 7, pp. 543-544, 558-559, 296; "The Skirmish at Sacramento, Kentucky," *Chicago* (Ill.) *Times*, Jan. 6, 1862, citing the *Louisville* (Ky.) *Democrat*; James Maynard Shanklin, *"Dearest Lizzie": Civil War Letters of James Maynard Shanklin*, Kenneth P. McCutchan, ed., Evansville, Ind., 1988, ppp. 141-142 note 4; Q.K. Juniper Wiggins, *Evansville* (Ind.) *Journal*, Feb. 13, 1862.

* * * * *

Albert Gallatin Bacon

Capt. Albert Gallatin Bacon lived for about an hour after Isaac and Elizabeth Johnson found him beside the road. Elizabeth Johnson brought him water from Truman Plain's spring. The Johnsons asked the captain if he was a citizen of the area. He replied that he was from Frankfort. They asked if he was a family man. He told them no, he was a bachelor. Isaac Johnson then asked if Bacon was a religious man. The soldier answered that he was not, then asked Isaac to pray for him. After Isaac prayed for Bacon, the mortally wounded captain prayed himself.

Bacon then asked Isaac Johnson if he could write, that he desired to have his will written. But Isaac, having no writing materials, could not comply with the captain's request. Isaac then asked Bacon if he did not live to write his will, or have it written, how he wanted his property disposed of. Bacon told the Johnsons that after all his just debts were paid, he willed his entire property to Sarah Ware Bacon, an unmarried sister living in Frankfort. According to the Johnsons, Albert Bacon was in "full possession of his faculties" and entirely rational. The captain lived another 25 or 30 minutes and "seemed to retain his faculties to the last moment."

On January 2, 1862, the Johnsons swore to the above in a statement before Presiding Judge D. Little, McLean Co. Court. On May 19, 1862, the Johnsons gave additional written testimony that the wishes of Captain Bacon had been "reduced to writing... within six days next after they were spoken to them." On June 28, 1862, their statements were probated in Franklin Co., Ky., court as the noncupative will of Albert G. Bacon.

As noted, Col. James Jackson buried the bodies of the killed enlisted men but not the body of Captain Bacon. The remains were sent to Frankfort by way of Louisville. 1st Lt. John J. Roberts, of Bacon's Company C, escorted the body home. Albert Bacon's funeral took place on January 2, 1862. According to a correspondent to the Louisville *Journal*, Bacon's was "the largest funeral cortege that (had) been seen in

Frankfort for years." Military honors were provided by the Frankfort
Home Guards, Lt. R.B. Taylor commanding. They were preceded in the
march from the Christian Church to the cemetery by Haley's Frankfort
Brass Band.

As with many Kentucky families, Albert Bacon had close relative on
both sides of the battle lines.His nephew, Albert Boult Fall, was an
artilleryman with Capt. Thomas K. Porter's Confederate Tennessee
Battery. But Albert also had a niece, Laura Bacon, who was married to
Union Maj. Eugene W. Crittenden, brother of Union Brig. Gen. Thomas
L. Crittenden and Confederate Brig. Gen. George Crittenden.

Sources: "The Affair at Sacramento," *Frankfort* (Ky.) *Yeoman,* Jan. 2, 1862; Ba-
con Military File, National Archives; Albert Boult Fall, "Civil War Letters of
Albert Boult Fall, Gunner For the Confederacy," *Register* of the Kentucky
Historical Society, Vol. 59, No. 2 (April 1961), pp. 150-168; pp. 150-157;
"The Late Capt. Bacon," *Louisville* (Ky.) *Daily Journal,* Jan. 3, 1862; *Louisville* (Ky.) *Daily Journal,* Jan. 4, 1862; Thomas Southard, "Sacramento sixth
graders report more local Civil War stories," *McLean Co. News,* Calhoun, Ky.,
Nov. 27, 1958; Ann Vance Todd, correspondence with author, August 7, 1994,
containing typed copies of depositions of Isaac and Elizabeth Johnson.

* * * * *

Edward Baker

Pvt. Edward Baker, Company B, 3d Kentucky, had been badly
wounded during the fight at Sacramento. The Confederates took him to
Greenville and there paroled and left him in the care of the citizens. R.T.
Martin, of Greenville, saw this prisoner the next day at Reno's Hotel.
Private Baker remained in the hotel for two months, and then went home
to Princeton.

In mid-January 1862, Confederate Col. John S. Scott led four
companies of his 1st Louisiana Cavalry on a scouting expedition to
Green River. They spent the night at Greenville, then went on to Roches-
ter where they "got" 65 hogs, then returned to their camp at Bowling
Green. Although his unit was not included in the patrol, Capt. Gus A.
Scott, Company E, 1st Louisiana, went along. While at a hotel in
Greenville he saw "a wounded Yankee, the first he had ever seen."
According to Captain Scott, the Union soldier had been "shot in several
places, in a fight at Sacramento" and, although unnamed, most likely was
Edward Baker. During the night, the Federal began "groaning terribly."
The Confederate captain went to him and found the wounded man had
become twisted in his bed and was unable to help himself. He made the
wounded man more comfortable and gave him some water. The "poor
fellow was grateful, but seemed very much surprised to think that a

Southern soldier would do *anything* for him."

Born in Coventry, England, Baker was discharged from the Union
Army on July 28, 1862. The surgeon's certificate states that Baker was
incapable of performing the duties of a soldier because of gun shot
wounds which had perforated his body and limbs. "Degree of disability,
entire!" (underlined by surgeon.) Following his discharge, Edward
remained in Princeton at least for a time. He worked for a time as a clerk
and did "such other light work as he could." On March 18, 1864 he
applied for an invalid pension. Although there is no firm statement in the
records, Baker apparently received the invalid pension. On January 1,
1868, Baker married Belle M. Lester in Lyon County, Kentucky.

On April 28, 1886, Mrs. Belle M. Baker applied for a widow's
pension for herself and four children based on the service of her recently
deceased husband, former Private Edward Baker, once of the 3d Kentucky Volunteer Calvary. This widow's pension was granted. She was
dropped from the roll because of her death on February 7, 1910. At the
time she was receiving $12 a month. This is the information so far
available. But an intriguing note appears in a report concerning an
incident that occurred in October 1864.

According to Charlton G. Duke, after August 1864 he, his brother
John C. Duke and a cousin, Capt. Lindsey Buckner, were captured in
west-central Kentucky while attempting to return to their command in
Tennessee. They were then sent to the prisoner-of-war prison at Louisville. They had been there some three or four weeks when they were
"coolly informed" that Capt. Lindsey Buckner, B.P. Wallace, and John
and Charlton Duke would be shot the following day by order of General
Burbridge in retaliation for a mail carrier who had been killed by a band
of guerrillas, supposed to be the Sue Mundy gang. "We did not spend
that day with any degree of pleasure, for the thought of dying such an
ignominious death at the hands of our enemies was indeed depressing,
but we determined to meet our fate like men," Charlton affirmed.

"We were greatly surprised the next morning when several nicely
dressed men in blue uniforms, one of whom we recognized as Mr. Ed
Baker, from our home at Princeton, Ky., came into the prison," Duke
continued. "He expressed pleasure at seeing me and my brother John,
and informed us that it was his great happiness to convey to us the good
news that through his influence and that of another prominent Union man
of Princeton General Burbridge had been persuaded to countermand the
order for our execution, and that we could have our choice of being sent
to a Northern prison or take the oath of allegiance and return home. We
thanked him and said we would go to prison. We asked if they could not

influence General Burbridge to release our companions also. He replied that he could do nothing for Captain Buckner but that Captain Wallace would probably be released, which was afterwards done."

Charlton sadly admitted that "a Captain Lirly, Lieutenant Blincoe, and on (sic) old man named Halley" were selected in their stead. "We little supposed that the men who had interfered in our behalf were actuated by any but kindly motives in securing our release from death, but soon found that those expressions of friendship had cost our mother $2,000 in cash, which she promptly forwarded to Louisville. Our friend Wallace was also ransomed by his friends, and had Captain Buckner's brother received in time the letter written him, his terrible fate would have been averted."

(This was obviously the October 25, 1864, execution, near Jeffersontown, Jefferson Co., of four men—Wilson P. Lilly, Company G, Confederate 1st Missouri Infantry; Capt. Lindsey Duke Buckner, officer in Col. J.Q.A. Chenoweth's Confederate cavalry and former member, Company H, Confederate 1st Kentucky Cavalry; Lt. William C. 'Dock' Blincoe, and 70-year-old Rev. Sherwood Hatley—in retaliation for the shooting, by guerrillas, of a Federal soldier. The four were taken to the spot of the killing by Capt. Rowland E. Hackett and 50 members of the Union 26th Kentucky Infantry and shot to death.)

This story would imply that Baker probably joined the Invalid Corps. This unit, officially the Veteran Reserve Corps, was composed of Federal soldiers who were unfit for full combat duty but could perform limited service. Some were used for guard duty. Baker may have been serving in this organization in October 1864, but so far no other information has been found showing such service by him. He may have, on the other hand, been acting in a civilian capacity.

Sources: Baker Military and Pension Files, National Archives; Mark M. Boatner III, *The Civil War Dictionary,* 4th printing, N.Y., 1966, "Veteran Reserve Corps"; Howell Carter, *A Cavalryman's Reminiscences of The Civil War,* 1900, rep. New Orleans, 1979, p. 21; Lewis Collins, and Richard H. Collins, *History of Kentucky,* 2 vols., Covington, Ky., 1882, Vol. 1, "Annals of Kentucky," under date of Oct. 25, 1864; E. Merton Coulter, *The Civil War and Readjustment in Kentucky,* Chapel Hill, N.C., 1926, p. 233; Charlton G. Duke, "Personal Prison Experiences and Death," *Confederate Veteran,* Vol. 19 (1911), p. 527; Robert Emmett McDowell, *City of Conflict,* Louisville, Ky., 1962, pp. 177-180; R.T. Martin, ""Recollections of the Civil War," Otto A. Rothert, *A History of Muhlenberg County,* Louisville, Ky., 1913, pp. 285-317, p. 297.

* * * * *

Charles Clark

In December 1861 Brig. Gen. Charles Clark was Lt. Col. Nathan Bedford Forrest's immediate commander. Born in Ohio, educated in Kentucky, Charles Clark went to Mississippi in 1831. He taught school, then practiced law and served as a state representative for several years. During the Mexican War he organized a company and later served as colonel of the 2d Mississippi Volunteers. After that war, Clark again served in the State House of Representatives from 1856-1861.

Initially against secession, by 1861 Clark had changed views and come to believe Mississippi's best course was to leave the Union. After the state's secession, he donated 100 bales of cotton to help arm the state's military forces. He formed a company of cavalry but was appointed a brigadier general in Mississippi's army. Then, in February 1861, the major general commanding the Army of Mississippi resigned. Jefferson Davis gave up the post to accept the presidency of the Confederacy. Gov. John Pettus appointed Charles Clark to replace Davis. General Clark held the position for only a few months before he too resigned, to accept a brigadier general's rank in the Confederate Army. In November 1861 General Clark was assigned command of a brigade, relieving Brig. Gen. Lloyd Tilghman at Hopkinsville, Ky.

After the Confederate withdrawal from Kentucky, they again met the Federals at Shiloh. General Clark was severely wounded in the shoulder. He recovered and fought in the Battle of Baton Rouge, La., where he led a division. Clark's hip was shattered by a minie ball and he was captured. He recovered but was permanently crippled and had to use crutches or a cane for the rest of his life.

photo from *Mississippi, Vol. VII, Confederate Military History*

After being held a prisoner in New Orleans, he was finally released February 21, 1863. He returned to Mississippi. A military hero on crutches, he was an ideal candidate for governor and was elected late in 1863. In spite of his crippling wounds and a long-standing bout with tuberculous, Clark remained a militant secessionist. In his inaugural address, he stated: "Rather than such base submission... let (us)... join hands together, march into the sea, and perish beneath the waters." But the war moved on to its inevitable end. In the spring of 1865 Governor Clark was arrested by Union authorities and imprisoned in Ft. Pulaski, Ga.

On September 12, 1865, the old fighter signed an oath of allegiance to the U. S. and returned to Mississippi. He practiced law and ran his plantation in Bolivar County. In 1876 he was appointed chancellor for the fourth judicial district, serving until his death on December 17, 1877. He was also a trustee of the University of Mississippi at the time of his death.

Sources: Mark M. Boatner III, *The Civil War Dictionary*, 4th printing, N.Y., 1966; Robert W. Dubay, "Mississippi Political, Civilian, and Military Realities of 1861: A Study in Frustration and Confusion," *Journal of Mississippi History*, Vol. XXXVI, No. 3 (Aug. 1974), pp. 215-242; Charles E. Hooker, *Mississippi*, Vol. VII, *Confederate Military History*, Clement A. Evans, ed., Atlanta, Ga., 1899, pp. 246-247; David G. Sansing, "Charles Clark Twenty-fourth Governor of Mississippi," *Mississippi History Now*, March 21, 2006 <http://mshistory.k12.ms.us/features/featured47/governors/19-charles-clark.html>; Jon L. Wakelyn, *Biographical Dictionary of the Confederacy*, Westport, Conn., 1977, p. 134; William F. Winter, "Mississippi's Civil War Governors," *Journal of Mississippi History*, Vol. LI, No. 2 (May 1989), pp. 77-88; Marcus Wright, "Memorandum of General Officers," in *Personnel of the Civil War*, Vol. 1, N.Y., 3d printing, 1968, pp. 207-376, pp. 235, 357.

* * * * *

Thomas L. Crittenden

Thomas L. Crittenden, Federal commander at Calhoun in December 1861, was a member of one of the most widely known divided families in Civil War Kentucky. Thomas Crittenden would rise to the rank of major general in the Union Army; his brother, George Bibb Crittenden, to major general in the Confederate Army.

Another interersting fact of his life concerns Thomas L. Crittenden and Jefferson Davis, president of the Confederacy. During the Mexican War they became friends. After a leave at home, Davis returned to his command, meeting up with then Col. Thomas Crittenden in Texas. Together they rode south thru the waste land of northern Mexico, "taking turns sleeping and watching," as one writer noted, "to avoid assassination at the hands of some lurking guerrillas." After Davis was wounded during the Battle of Buena Vista, it was Crittenden Davis asked to write to his brother Joseph Davis with information on his wound and the bravery of his unit, the Mississippi Riflemen, during the battle.

Although often called a political general, Thomas Crittenden distinguished himself during the war, especially at Shiloh, Stone River and Chickamauga. He was promoted major general of volunteers following the Battle of Shiloh. At Stone River, Tenn., he was highly praised for gallantry and received a brevet promotion to brigadier general in the regular army. At the Battle of Chickamauga, however, his command was routed by the Confederates. Later he was transferred to the East where he commanded the 1st Division, IX Corps, until December 13, 1864, when he resigned.

After the war, Thomas Crittenden joined the regular U.S. Army and, on July 28, 1866, was appointed colonel, commanding 32d Infantry Regiment. He was transferred to command of the 17th Infantry in 1869. For the next dozen years he served at several posts, while remaining commander of the 17th, until he retired May 19, 1881, to live on Staten Island. He died in Annandale, N.Y., October 23, 1893.

One of Colonel Crittenden's saddest times came with the loss of his only son, 2d Lt. John J. Crittenden. In the fall of 1875 John Crittenden was appointed second lieutenant in the 20th Infantry Regiment. Shortly afterward, a shotgun cartridge exploded in his face, causing him to lose his left eye. The following year, although still recuperating, the young officer learned about a cavalry operation being undertaken. Col. George A. Custer agreed to let him fill one of several vacancies in the 7th Cavalry. At the Battle of the Little Big Horn, June 25, 1876, 22-year-old Lieutenant Crittenden served as second-in-command to 1st Lt. James Calhoun in L Troop. Lieutenant Calhoun was Colonel Custer's brother-in-law, being married to the colonel's sister Margaret. (During the Battle of the Little Big Horn, the Custer family was well represented. In addition to George A. Custer and his brother-in-law James Calhoun, George's brother, Capt. Thomas Custer, Company C; his civilian brother, 27-year-old Boston Custer and civilian nephew, Harry 'Autie' Reed, both accompanying Colonel Custer at the latter's suggestion, died at the Little Big Horn.)

Lieutenant Crittenden had been told by his father to do his duty and never retreat. The two L Troop officers are believed to have died gallantly, trying to hold their skirmish line together. They were overwhelmed on what is now known as Calhoun Hill. Lieutenant Crittenden was identified by his shattered glass eye. Col. Thomas Crittenden insisted his son be buried where he had fallen. The remains of all other identified offcers were removed in 1877. In 1932, Lieutenant Crittenden's body was also removed to the main battle cemetery. The lieutenant's watch was recovered from the Indians some time after the battle and returned to the family in 1880.

Sources: Mark M. Boatner III, *The Civil War Dictionary,* 4th printing, N.Y., 1966; Crittenden Military File, National Archives; H.H. Crittenden, comp., *The Crittenden Memoirs,* N.Y., 1936, p. 508; Helen Bartter Crocker, "A War Divides Green River County," *Register* of the Kentucky Historical Society, Vol. 70, No. 4 (Oct. 1972), pp. 295-311, pp. 307-308; Trevor N. Dupuy, Curt Johnson and David L. Bongard, *Harper Encyclopedia of Military Biography*, Edison, N.J., 1992; Webb Garrison, *A Treasury of Civil War Tales,* Nashville, Tenn.,

1988, Chapter 8: "George and Thomas Crittenden Mirrored the Sundered Nation," pp. 41-44; John T. Hubbell and James W. Geary, eds., *Biographical Dictionary of the Union*, Westport, Conn., 1959, pp. 118-119; Neil C. Mangum, "The General's Tour: The Little Bighorn Campaign: Civil War Veterans Die on the Plains," *Blue & Gray Magazine*, Vol. 23, No. 2, 2006, pp. 6-27, 42-65, pp. 21, 22, 45; Charles M. Robinson III, *A Good Year To Die: The Story of the Great Sioux War*, N.Y., 1995, p. 192; Douglas D Scott, Richard A. Fox Jr.; Melissa A. Connor and Dick Harmon, *Archaeological Perspectives on the Battle of the Little Big Horn*, Norman, Okla., 1989, pp. 13, 93, 246, 248; Douglas D. Scott, P. Willey and Melissa A. Connor, *They Died with Custer: Soldiers' Bones from the Battle of the Little Bighorn*, Norman, Okla., 1998, scattered references; Henry E. Simmons, comp., *A Concise Encyclopedia of the Civil War*, (New York, 1965), p. 64; Larry Sklenar, *To Hell with Honor: Custer and the Little Bighorn*, Norman, Okla., 2000, scattered references; Hudson Strode, *Jefferson Davis American Patriot 1808 - 1861*, N.Y., 1955, pp. 175 - 185; William O. Taylor, *With Custer on the Little Bighorn*, N.Y., 1996, Appendix II, pp. 198-199.

* * * * *

Arthur N. Davis

Capt. Arthur N. Davis, 3d Kentucky Cavalry, dislocated his shoulder and was captured at Sacramento. After being taken to Hopkinsville, Captain Davis was taken to Nashville, Tenn., then to the Salisbury, N.C., Prisoner-of-War camp. From there he was sent to Libby Prison. He was held prisoner by the Confederates until September 21, 1862. At that time, an exchange of prisoners of war took place at Aiken's Landing, Va., with Captain Davis being exchanged for Capt. W.W. Morris, 42d Virginia Infantry.

Captain Davis' injured right shoulder was never properly set and did not heal correctly, eventually ending his military career. On April 29, 1863, Captain Davis resigned on a surgeon's certificate of disability. The surgeon stated that his injury did not wholly disqualify him for military service, but it did render any action with the saber impossible. On August 23, 1866, Arthur Davis applied for an invalid's pension stating that he was unable to do manual labor and further stating that he was "one-half disabled from obtaining his subsistence from manual labor in consequence of his above named injuries."

A Tennessee native, Davis had come to Kentucky in 1847 after serving in the Mexican War. He was killed in Muhlenberg Co. on March 9, 1868, by the falling of a tree branch. He was buried on his farm in Muhlenberg Co. (Otto Rothert states Davis' death occurred about 1872 and that a bough of a tree fell on the carriage in which Davis and his family were riding, killing only the former captain.)

Sources: Davis Military and Pension Files, National Archives; Sworn statement of George B. Eades, Greenville, Ky., postmaster dated Aug. 11, 1869, in Davis Pension files; Otto A. Rothert, *A History of Muhlenberg County,* Louisville, Ky., 1913. pp. 192; *The War of the Rebellion: A Compilation of the Official Records of the Union and Confederate Armies,* Ser. II, Vol. 4, Washington, D.C., 1880-1901. Series II, Vol. 3, pp. 418-419, Vol. 4, p. 578; *3d Ky. Cav. Memo. Book,* National Archives, pages unnumbered

* * * * *

Nathan Bedford Forrest

Following Sacramento the war moved on, gaining in fierceness. Nathan Bedford Forrest, the Confederate commander at the Battle of Sacramento, carved out a niche in history as one of the greatest natural-born cavalrymen—and perhaps military leaders—ever produced in America. A remarkable prediction had been made almost two months before the Sacramento fight. In a letter, dated November 4, 1861, to Southern Gen. Albert S. Johnston, commander of Confederate Armies west of the Appalachians, Sam Tate wrote: "Give Forrest a chance, and he will distinguish himself." Sacramento started his rise to military fame.

Forrest's fame increased greatly in February 1862 when he received permission to withdraw from Ft. Donelson before the official surrender. He took out not only most of his own troopers but many others who decided to follow him out rather than go into a Union prison camp. Following that, just before Shiloh, Forrest's battalion was increased to a regiment—officially the 3d Tennessee Cavalry but better known as Forrest's Old Regiment—and he was promoted to colonel. In June 1862, after masterful rear guard duties during the withdrawal from Shiloh, he assumed command of a cavalry brigade.

On July 21, 1862, after he captured the Federal garrison at Murfreesboro with its stores, Forrest was promoted to brigadier general. He was promoted to major general December 4, 1863. By then his fame as a cavalry leader was almost legendary. "There was never a man like Forrest," retired Maj. Gen. John K. Herr, the U.S. Army's last Chief of Cavalry wrote. "Since the days of Richard Coeur de Lion (Richard the

photo from Jordan and Pryor's *Campaigns of Lieut.-Gen. N.B. Forrest*

Lionhearted) there has seldom appeared a man who so fired the enthusiasm of men by his primitive lust for personal combat combined with
brilliant leadership and planning." Union Gen. U.S. Grant described him
as "about the ablest cavalry general in the South." It might, in fact, be
said of Forrest, as it was said of the Roman general and dictator Sulla,
"that he was half lion and half fox, and that the fox in him was more
dangerous than the lion." Forrest, without benefit of military education,
exemplified Napoleon I's maxim that "In war the general alone can judge
of certain arrangements. It depends on him alone to conquer difficulties
by his own superior talents and resolution."

In June 1864 he routed a superior force at Brices Cross Roads and
the following month stood off Gen. A.J. Smith at Tupelo. But before
those brilliant victories came the controversial capture of Ft. Pillow. Ft.
Pillow is still debated and argued. Was there a massacre of black Union
troops, and if so, to what extent was Forrest to blame? Was he actually
on the scene when some black soldiers were apparently killed after they
surrendered? As Paul Ashdown and Edward Caudill note, in *The Myth of
Nathan Bedford Forrest,* what happened at Fort Pillow is "one of the
most studied and highly contentious episodes of the entire Civil War."

But the war moved on. In November and December 1864 General
Forrest served under Gen. John Bell Hood during the Tennessee Campaign, commanding all Confederate cavalry. After Hood's defeat Forrest
again conducted a brilliant rearguard action which allowed the Confederate forces to withdraw southward. Forrest was promoted to lieutenant
general on February 28, 1865. He was finally overwhelmed by greatly
superior forces at Selma, Ala., in April 1865. On May 9, 1865, Forrest
gave his farewell to his troops:

"That we are beaten is a self-evident fact, and any further resistance
on our part would be justly regarded as the very height of folly and
rashness... Reason dictates and humanity demands that no more blood be
shed... Civil War, such as you have just passed through, naturally engen-
ders feelings of animosity, hatred and revenge. It is our duty to diverst
ourselves of all such feelings, and, so far as it is in our power to do so, to
cultivate friendly feelings toward those with whom we have so long
contested, and heretofore so widely but honestly differed...You have been
good soldiers, you can be good citizens. Obey the laws, preserve your
honor, and the govenment to which you have surrendered can afford to
be and will be magnanimous."

After the war Forrest, like many Southerners, was broke. He again
took up the life of a planter and also became involved in railroad promotion. Then came the next great controversy in his life: his role in the

early Ku Klux Klan. The best evidence suggests Forrest was active in the Klan and most likely served as its first Grand Wizard. It should be noted, however, that the original Klan arose in the desperate year of 1867 and was officially disbanded in 1869, when it became evident that it could not be effectively controlled by its more responsible members. It activities in this short time were not all terrorism and violence, and in many sections its aims were to maintain order and fill the void created by the collapse of local law enforcement. It also opposed a similar organization called the Loyal League which preyed on former pro-Confederates under the protection of Tennessee Governor Parson Brownlow. "Abolish the Loyal League and the Ku Klux Klan; let us come together and stand together." Bedford Forrest announced, calling for an end to the civil unrest. His plea was echoed by other former Confederate leaders, including Frank Cheatham, Bushrod Johnson, George Maney, William Bate and Gideon Pillow.

As to his racism, Bedford Forrest was far from the only former Confederate who changed his views about the Black man following the war. He had been a slave trader and slave owner in his younger life, but by mid-1875 his attitude had undergone a major shift. On July 5, 1875, he was invited to address, and did speak, to the Independent Order of Pole-Bearers Association, an organization formed to promote the economic and civil rights of Negroes. At the Memphis Fair Grounds Forrest said. "We were born on the same soil, breathe the same air, live on the same land, and why should we not be brothers and sisters?" He also told them: "When I can serve you I will do so. We have but one flag, one country; let us stand together. We may differ in color, but not in sentiment." Abraham Lincoln has been justly praised for a growth in character during the war in his attitude toward slavery and Blacks. Are others not to be given the same considerations? Is Forrest denied the right to change?

About 1876 Forrest visited his legal adviser, Gen. John T. Morgan, and ordered him to discontinue all litigation with which he was connected. "General," Forrest told him, "I am broken in health and in spirit and have not long to live. My life has been a battle from the start... I have seen too much violence, and I want to close my days at peace with all the world, as I am now at peace with my Maker."

In early summer 1877 Maj. Charles W. Anderson, a long-time friend came to visit Forrest. At that time Forrest was bedridden at his brother Jesse's home. Anderson was stunned by the change in his old commander. "Major, I am not the same man you were with so long and knew so well," Forrest said, noticing Anderson's reaction. "I hope I am a better man now than then. I have been and am trying to lead another kind of life. Mary (Bedford's wife) has been praying for me night and day for all these years, and I feel now that through her prayers my life has been spared and I have passed safely through so many dangers." The often profane and violent Forrest had been converted by his wife's prayers two years before and joined her Cumberland Presbyterian Church.

Former Confederate Pres. Jefferson Davis happened to be in Memphis in 1877. He heard Forrest was gravely ill and hurried to his bedside. Bedford roused enough to speak to him. Nathan Bedford Forrest died the following day, in the evening of Monday, October 29, 1877, possibly of diabetes. (An interesting note here is that both the Confederate and Union commander at the Battle of Sacramento probably died of diabetes.) Among those present at the bedside of Forrest when he died was a young Lee Meriwether, son of a friend of Forrest's. "That was the first time I ever saw Death," Lee later wrote. "I had always thought of Death as a cruel monster, but he entered that room so silently, so stealthily, Death seemed almost a friendly visitor. There was no struggle, no pain, Forrest simply ceased to breath as he sank into the vast ocean of yesteryears." (Lee was apparently no relation to Ned Meriwether who died at Sacramento.)

Jefferson Davis was a pallbearer. Riding in a carriage behind the body to the cemetery, he praised Forrest highly to Tennessee Gov. James D. Porter. But Davis confessed he saw Forrest's full worth as a commander only after his fall 1864 campaign. Davis admitted he had been "misled" earlier by "generals" who rated Forrest only a "partisan raider." Military history has truly shown that Nathan Bedford Forrest was a much greater soldier than a mere "partisan raider."

(Forrest's grandson, Brig. Gen. N.B. Forrest Jr., a 1928 West Point graduate, was shot down leading an air attack over Germany during World War II, in 1943, at the age of 38.)

Sources: Felicity Allen, *Jefferson Davis: Unconquerable Heart,* Columbia, Mo., 1999, p. 507; Paul Ashdown and Edward Caudill, *The Myth of Nathan Bedford Forrest,* N.Y., 2005, pp. 11, 31-39; Gerald Astor with Joseph Millard, "Heller on Horseback," in Joseph Millard, ed., *True Civil War Stories,* Greenwich, Conn., 1961, pp. 169-184; Randall Bedwell, ed., *May I Quote You, General Forrest?,* Nashville, Tenn., 1997, pp. 76, 79; Martin Blumenson and James L. Stokesbury, *Masters of the Art of Command,* 1975, rep. N.Y., n.d., p. 137; Robert M. Browning Jr., *Forrest: The Confederacy's Relentless Warrior,* Washington, D.C., 2004; Trevor N. Dupuy, Curt Johnson and David L. Bongard, *Harper Encyclopedia of Military Biography,* Edison, N.J., 1992; "General Nathan Bedford Forrest: War Hero or Villain?," *Armchair General,* November 2005, p. 23; Mark Grimsley, "Millionaire Rebel Raider: The Life of Nathan Bedford Forrest," part 1, *Civil War Times Illustrated,* Oct. 1993, pp. 58-73; Ashley Halsey Jr., *Who Fired The First Shot?,*N.Y., 1963, "What They

Fought With," pp. 48-81, p. 54; James Hamilton, *The Battle of Fort Donelson,*
N.Y., 1968, numerous mentions; John K. Herr and Edward S. Wallace, *The Story
of the U.S. Cavalry 1775-1942,* Boston, Mass., 1953., p. 100; Jack Hurst,
Nathan Bedford Forrest: A Biography, N.Y., 1993; Jack Hurst, *Men of Fire:
Grant, Forrest, and the Campaign That Decided the Civil War,* N.Y., 2007;
Thomas Jordan and J.P. Pryor, *The Campaigns of Lieut.-Gen. N.B. Forrest, and
of Forrest's Cavalry,* 1868, rep. Dayton, Ohio, 1973; Andrew Nelson Lytle,
Bedford Forrest And His Critter Company, N.Y., 1931; William D. McCain,
"Nathan Bedford Forrest: An Evaluation," *Journal of Mississippi History,* Vol.
24, No. 4 (Oct. 1962), pp 203-225; Lee Meriwether, *My First 98 Years 1862-
1960,* Columbia, Mo., 1960, pp. 22-25. Jerry D. Morelock, "Hell To Pay:
Bedford Forrest At Brice's Crossroads," *Armchair General,* November 2005, pp.
44-49; Napoleon I, *The Military Maxims of Napoleon,* Lt. Gen. Sir George C.
D'Aguilar, David G. Chandler, commentary, 1831, rep. Mechanicsburg, Pa.,
2002, p. 77; Paul E. Steiner, *Medical-Military Portraits of Union and Confederate Generals,* Philadelphia, 1968; William J. Stier, "Fury Takes The Field," *Civil
War Times Illustrated,* December 1999, pp. 40-48; *War of the Rebellion: A
Compilation of the Official Records of the Union and Confederate Armies*
(commonly *Official Records of the Union and Confederate Armies*), 128
vols., index, and atlases, Washington, D.C., 1880-1901, Ser. I, Vol. 49, Pt. 2,
pp. 1289-1290; Brian Steel Wills, *A Battle from the Start: The Life of Nathan
Bedford Forrest,* N.Y., 1992; Joe Weller, "Nathan Bedford Forrest: An Analysis
of Untutored Military Genius," *Tennessee Historical Quarterly,* Vol. 18, No. 3
(Sept. 1959), pp. 213-251; Viscount Garnet Wolseley, "General Viscount
Wolseley on Forrest," in Robert Selph Henry, ed., *as they saw Forrest,* Jackson,
Tenn., 1956, pp. 17-53; "Publisher's Note," in John Allan Wyeth, *That Devil
Forrest,* 1899, reprint Baton Rouge, La., 1989, pp. xi-xiv, pp. xi-xii.

* * * * *

Alvin Fowler

Alvin 'Al' Fowler is alleged to have been the Confederate who
attacked Pvt. John L. Williams after the latter was unhorsed and unarmed
at the Battle of Sacramento. This allegation ignores the fact that in
Greenville immediately after the battle the report was that "some of
(Forrest's) men rushed up and began using sabers on Williams." Note
that the report states men, plural. Fowler may, however, have served with
Forrest at the time, as an unenlisted volunteer.

Fowler was born in Hopkins County, Kentucky, on July 17, 1835.
Even before enlisting in the Southern Army Fowler served the South. On
September 15, 1861, Union recruits, under Col. James F. Buckner,
escaped from their camp near Hopkinsville as Confederates moved into
the town. That night Colonel Buckner and 25 of his men camped in a
frame Baptist Church at Burnt Mill, Webster County, Kentucky. Fowler
gathered pro-Southern friends and during the night surrounded the

church. At daybreak the Southerners attacked. The Federals surrendered after an hour-long skirmish. One Southerner, Buck Madison, was slightly wounded. The Federals were all captured, alive and unwounded.

Reportedly, Fowler served under Nathan Bedford Forrest as a private at Sacramento and later at Ft. Donelson, Tenn. He is said to have "endeared himself" to Forrest "by his indomitable courage." Later in 1862 Al Fowler joined Col. Adam R. Johnson (see later bio) in raising the Confederate 10th Kentucky

photo from Adam R. Johnson's Partisan Rangers of the Confederate Army

Partisan Rangers. Fowler was appointed captain of Company I.

Colonel Johnson scattered his men to give the impression his strength was greater than it actually was. Fowler became notorious throughout the counties along the lower Green River. Hopkins and Muhlenberg Counties especially knew his raiding. Union officials constantly sought to capture the illusive captain and his men. They constantly failed. In one reported incident, they attempted to capture Fowler in Madisonville. Although they knew he was there, they did not know what he looked like. Being out of uniform, he audaciously helped Union officers place their pickets. That night he easily and safely made his way out of town.

On November 5, 1862, near Sumner's Store in Muhlenberg Co., Captain Fowler led his men in an attempt to slip up on the encamped 1st Battalion, Union 8th Kentucky Cavalry led by Maj. James H. Holloway. The Federals were alert. When the Southerners fired their first shot, Union troops immediately returned fire. The Confederate captain ordered his men to lie down and return fire. Apparently Fowler was killed almost immediately. Some reports state Fowler was hit from behind, apparently accidentally killed by his own men. But a Union soldier later reported that Fowler was hit by a pistol ball in the head and two carbine balls in the chest, all in the front. Two other Confederates were killed, 16 captured and a large number wounded. The Southern force was routed. (Many sources give the Union commander's rank as colonel, a rank he later held; many sources also give Fowler's rank as colonel, a rank he never held. Some later reports also give the date as November 23 and some give the site's location as in Hopkins County.)

Sources: James Duane Bolin, *An Abiding Faith: A Sesquicentennial History of
Providence, Kentucky, 1840-1990*, Providence, Ky., 1990, p. 17; L.D.
Hockersmith, *Morgan's Escape: A Thrilling Story Of War Times*, 1903, rep.
n.d., n.p., p. 6; Polk Laffoon, "Al Fowler," in Adam R. Johnson, *Partisan
Rangers of the Confederate States Army*, William J. Davis, ed., 1904, rep.
Evansville, Ind., n.d., pp. 300-303; Richard T. Martin, "Recollections of the
Civil War," Otto A. Rothert, *A History of Muhlenberg County*, Louisville,
1913, pp. 285-317, pp. 299-300; Frank Amplias Owen, "An Outline History
of the Partisan Rangers," in Adam R. Johnson, *Partisan Rangers of the
Confederate States Army*, William J. Davis, ed., 1904, rep. Evansville, Ind.,
n.d., pp. 344-352, pp. 350-351; Union Soldiers and Sailors Monument
Association, *Union Regiments of Kentucky*, Louisville, Ky., 1897, pp. 273-
274; Wallace M. Wadlington and David M. Sullivan, "Our Town: A History
of Madisonville, Kentucky," in *Just The Other Day: A History of
Madisonville, Kentucky*, J.T. Gooch, ed., Madisonville, Ky., 1981, pp. 8-26,
p. 17.

* * * * *

Nicholas C. Gould

Within six weeks after the skirmish at Sacramento, Capt. Nicholas C.
Gould and his company were part of Lt. Col. Nathan Bedford Forrest's
cavalry battalion operating at Ft. Donelson, Tenn. When Forrest with-
drew from the fort before the Confederate surrender, most or all of
Captain Gould's company failed to accompany their colonel out. "Why it
(Gould's company) did not come out was never explained," according to
Jordan and Pryor.

An obituary of Trooper Jack O'Donnell, a member of Gould's
company, suggests that only part of the company failed to withdraw from
Ft. Donelson. This reports that "a portion" of the company under Lt.
M.L. Sims surrendered and was sent to Camp Butler. After six months
they were exchanged at Vicksburg, Miss.Here they organized a new
company with Sims as captain and then went to Texas and joined the 23d
Texas Cavalry under Colonel Gould.

Fitzhugh implies that Captain Gould was with Forrest at Corinth,
Miss., following the Battle of Shiloh in early April 1862. He writes that
at Corinth the Texas company, Company D, of Forrest's command left
Forrest and returned to Texas. At Clarksville, Red River Co., Texas,
Nicholas Gould, now with a commission to raise a regiment of cavalry,
disbanded his company and organized the 23d Texas Cavalry Regiment.
The organization of the 23d officially began about April 12, 1862, but
was not completed until October 25, 1862, with 10 companies, A through
K. Most of his old company joined the new regiment, many as officers,
although offically Forrest's former Company D now served as the 23d's

Company K. Gould served as colonel commanding the 23d.

The 23d Texas Cavalry served in Texas and Indian Territory (now Oklahoma) until March 1864. They then joined Maj. Gen. Richard Taylor's command in western Louisiana. Here they were engaged in the pivotal battles of Mansfield, on April 8, 1864, and Pleasant Hill, on April 9. These battles helped drive back Federal forces during the Red River Campaign. In February 1865 the 23d was dismounted. They were surrendered by Gen. E. Kirby Smith, commanding the Confederate Trans-Mississippi Department, on May 26, 1865.

Nicholas C. Gould, a pre-war attorney in Red River County, apparently returned there after his surrender. His wife and children appear in the 1870 census for Red River County, but he does not. He seems to have died in the five years following the war. Gould is buried in the Clarksville Baptist Cemetery, Red River County, Tex. Former Lt. Col. John A. Cotley of the 23d is buried next to him.

Sources: Donaly E. Brice, Sr. Research Asst., Texas State Library and Archives Commission, correspondence, Dec. 21, 2007, Lester N. Fitzhugh, "One Texas Unit Served Forrest," unidentified newspaper clipping.; Harry McCorry Henderson, *Texas in the Confederacy,* San Antonio, Tex., 1955, p. 130; "Jack O'Donnell," *Confederate Veteran,* Vol. 13 (1905), p. 175; Thomas Jordan and J.P. Pryor, *The Campaigns of Lieut.-Gen. N.B. Forrest, and of Forrest's Cavalry,* 1868, rep. Dayton, Ohio, 1973, footnote, p. 93; James A. Mundie Jr., Dean E. Letzring, Bruce S. Allardice and John H. Luckey, *Texas Burial Sites of Civil War Notables,* Hillsboro, Tex., 2002, p. 113; Carl A. Russell, "Chronology of Reported Events, Gould's 23 Regiment," "In memory of Thomas H. Massie, Private, Company B, Gould's Regiment, 23rd Texas Cavalry, C.S.A., p. 1; Stewart Sifakis, *Compendium of the Confederate Armies: Texas,* N.Y., 1995, p. 80; William J. Stier, "Fury Takes The Field," *Civil War Times Illustrated,* December 1999, pp. 40-48., p. 44; Dudley G. Wooten, ed., *A Comprehensive History of Texas, 1685 to 1897,* Dallas, Tex., 1898, p. 636; Marcus J. Wright, *Texas in the War 1861-1865,* Harold B. Simpson, ed., Hillsboro, Tex., pp. 27, 119 n. 199.

* * * * *

James M. Holmes

Capt. James M. Holmes, Company B, 3d Kentucky Cavalry, who led the advance of the relief force to Sacramento, received a promotion shortly after the Battle of Sacramento. The captain was promoted to major January 10, 1862, to fill the vacancy created by the resignation of Maj. Green Clay. Seven months later, on August 13, 1862, Holmes was promoted to lieutenant colonel. He held the rank for less than a year before being forced to resign, effective May 27, 1863. He had, in fact, been on detached duty from December 20, 1862, until February, being unfit for field duty. He was again on duty with the 3d Kentucky through April but was on court martial duty during early May. Colonel Holmes had contracted a fever which affected the optic nerves. His eyesight slowly deteriorated. On May 16, 1863, he tendered his resignation, accompanied by a surgeon's certificate of disability.

Colonel Holmes remained at home until February, 1864. On the recommendations of officers with whom he had served, he was commissioned captain and brevet lieutenant colonel in the regular Army. He reentered service and was sent to Connecticut to muster out the volunteer troops from that state. In January 1866 he resigned his commission and returned to Daviess Co., Ky. and his 470-acre farm.

He was an active prohibitionist and, in 1891. was nominated by the Prohibition Party for state treasurer. He had reportedly been the first person in Daviess Co. to cast a Prohibitionist vote. When he started campaigning for the office of state treasurer, Holmes gave up farming for good, although his campaign ended in failure.

Born November 10, 1825, James M. Holmes lived until a few months short of his 100th birthday, dying July 18, 1925. His mind had remained active until a short time before his death, although he had been unable to move about without help for several months.

Sources: "Col. J.M. Holmes Dies Saturday of Infirmities," *Owensboro* (Ky.) *Messenger,* July 19, 1925; William Foster Hayes, *Sixty Years of Owensboro,* Owensboro, 1943, pp. 26-27; *History of Daviess County, Kentucky,* 1883, reprint Evansville, Ind., 1966, pp. 735-736; Holmes Military Files, National Archives; Kentucky Adjutant General, *Report of the Adjutant General of Kentucky, 1861-1866,* 2 vols., Frankfort, Ky., 1866-1867, Vol. 1, p. 68, Vol. 2, p. 930; Hambleton Tapp and James C. Klotter, *Kentucky: Decades of Discord: 1865-1900,* Frankfort, Ky., 1977, p. 321.

* * * * *

James Streshly Jackson

Col. James S. Jackson worked hard to mold his 3d Kentucky Cavalry into a disciplined, efficient force. A Louisville correspondent observed the unit in training in mid-February 1862. "We witnessed this morning a drill by Jackson's cavalry regiment," Clarence Linden wrote for the *Louisville* (Ky.) *Journal* on February 13, 1862, "and never have our eyes gazed upon a more animated scene. The sloping hills, which are now beaten down to the smoothiness of a good road, afford them ample scope for their field maneuvering."

Colonel Jackson, according to Linden, was "the idol of his soldiers." The correspondent wrote of Jackson that a "more polished, urbane, eloquent, brave, and qualified officer is not in our army." He stressed that Jackson gave his men "full assurance that on the battlefield he will lead them to victory." Perhaps the reporter had an inkling of the future when he added: "We think him destined, should he survive the dangers of the campaign, to take a high place in the councils of the nation."

Soon after the drill observed by Linden, the 3d Kentucky and Behr's Indiana Artillery left Calhoun and moved overland south toward the front. First to Bowling Green, then Nashville and on to Pittsburgh Landing they rode. After Shiloh, as the Union Army moved on toward Corinth, Colonel Jackson was placed in command of the Army of the Ohio's cavalry brigade. On August 13, 1862, he was promoted to brigadier general of volunteers.

When Gen. Braxton Bragg moved into Kentucky later in August 1862, General Jackson and his brigade were at Decherd, Tenn. From there Jackson led his brigade back into Kentucky in a race to Louisville, a race the Federals won. Jackson's good friend, Maj. Gen. William 'Bull' Nelson, commanding the Union Army of Kentucky, had been organizing the defense of Louisville. When Maj. Gen. Don Carlos Buell arrived, he planned to rely on Nelson as the Federals attempted to drive the Confederates out of the state. Then tragedy struck.

In a still controversial, tragic affair, Union Brig. Gen. Jefferson C. Davis shot and fatally wounded an unarmed General Nelson. General Jackson and Brig. Gen. William Terrill were hard to appease, the latter calling for the immediate hanging of Davis. (Maj. Gen. Thomas L. Crittenden, another of Nelson's friends, was one of the first to reach the dying general.) Although Davis was held in arrest in Louisville for a short time, the necessity to move against the Confederates prevented any other immediate response to the killing. (Because Buell was shortly afterward removed from command and Jackson and Terrell were no longer available to push an investigation, Davis was never punished for the killing.)

As the Union forces moved out of Louisville, General Jackson commanded the Tenth Division, in Maj. Gen. Alexander McD. McCook's I Corps. The division was heavily engaged in the Battle of Perryville, Ky., on October 8, 1862. As Confederates moved to roll up the Federal left, and the Federal left began to give, General Jackson stood near Parson's Battery. He watched as Southerners climbed a fence, advancing, firing steadily all the while. Capt. Samuel M. Starling, Tenth Division's inspector general, stood within a few feet of Jackson. He later stated that, with the "minnie balls.. with their peculiar whizzing noise" coming in greater and greater numbers, Jackson remarked, "Well, I'll be damned if this is not getting rather particular." Almost as soon as he finished speaking, two shots struck him in the right breast. He struggled to speak but was unable and died within a few minutes.

The night before the Battle of Perryville, according to Col. Charles Denby, Union 42d Indiana Infantry (which had been at Calhoun in 1861), Generals Jackson and Terrell and Col. George Webster discussed the chances of being hit by fire during an engagement. They agreed that men would not be frightened if they considered the probabilities and how slight the chance of any particular person being killed. While their theory might have been sound in most cases, in their cases it failed. All three were killed at Perryville on October 8. (General Terrill was another example of brother against brother. From Virginia, his brother died a Confederate colonel leading his troops in the East.)

After his death James Jackson's body was moved to the rear of the Open Hill to await an ambulance. Before two staff officers could bring an ambulance back, Confederates overran the area where Jackson lay. When the two officers returned, they found "quite a number of Rebel & Union soldiers, ministering to the wounded and looking at the [battle]ground... We found Jackson just where we left him, laid out very straight, with his boots, hat & buttons taken, but his body untouched. We put him in an ambulance & dashed on towards Louisville." His body was then placed in a vault in Cave Hill Cemetery. On March 24, 1863, his remains were reinterred in the cemetery at Hopkinsville.

Sources: Mark M. Boatner III, *The Civil War Dictionary,* 4th printing, N.Y., 1966; James B. Fry, *Killed by a Brother Soldier,* N.Y., 1885; Charles C. Gilbert, "On The Field of Perryville," in Robert Underwood Johnson and Clarence Clough Buel, eds., *Battles and Leaders of the Civil War,* 4 Vols., 1887 - 1888, reprint N.Y., 1956, Vol. 3, pp. 52-59, footnote, p. 57; Arthur A. Griese, "A Louisville Tragedy—1862," *Filson Club History Quarterly,* Vol. 26, No. 2 (April 1952), pp. 133-154; Kenneth A. Hafendorfer, *Perryville: Battle for Kentucky,* Louisville, Ky., 1991, p. 216; John T. Hubbell, and James W. Geary, eds., *Biographical Dictionary of the Union,* Westport, Conn., 1959, pp. 273-274; Joyce, John A., *A Checkered Life,* Chicago, 1883, p. 73; Clarence Linden, "Particulars of the Sacramento Skirmish," *Louisville* (Ky.) *Journal,* Feb. 24, 1862; James Lee McDonough, *War in Kentucky: From Shiloh to Perryville,* Knoxville, Tenn., 1994, paper reprint 1996, pp. 252-253; Charles Mayfield Meacham, *A History of Christian County, Kentucky from Oxcart to Airplane,* Nashville, Tenn.,

1930, pp.145-146; William Henry Perrin, ed., *Counties of Christian and Trigg,* Chicago & Louisville, 1884, pp. 188-189; Stephen Z.Starr, *The Union Cavalry in the Civil War:* Vol. I: *From Fort Sumter to Gettysburg 1861-1863,* Baton Rouge, La., 1979, p. 83 n. 53; Richard G. Stone Jr., *Kentucky Fighting Men 1861-1945,* Lexington, Ky., 1982, pp. 15-16; Union Soldiers and Sailors Monument Association, *The Union Regiments of Kentucky,* Louisville, 1897, pp. 63, 133-135; *War of the Rebellion: A Compilation of the Official Records of the Union and Confederate Armies* (commonly *Official Records of the Union and Confederate Armies*), 128 vols., index, and atlases, Washington, D.C., 1880-1901, Ser. I, Vol. 7; Ser. II, Vol. 3, Vol. 4.

* * * * *

Calvin N. Jarrell

The actions of 1st Sgt. Calvin N. Jarrell, Company D, at Sacramento may have resulted in his promotion, although there is no direct evident. The 27-year-old joined the 3d Kentucky Cavalry as 1st Sergeant on October 15, 1861. On January 26, 1862, less than a month after the Battle of Sacramento, he was promoted to 2d lieutenant in Company D. Unfortunately, he was able to serve as an officer for less than four months. On May 4, 1862, he resigned due to ill health. His resignation request was forwarded with approval by Col. James S. Jackson, who commented that Lieutenant Jarrel had been "of no service to the regiment for some time" because of chonic disease in the back.

Born in Tennessee, Jarrell and his family moved to Kentucky during the 1850s. In 1872 they moved to Texas. Either there or before he left Kentucky, Calvin Jarrell became a Cumberland Presbyterian preacher. Rev. W.P. Kloster wrote that Reverend Jarrell "did not wait to have a call from some well organized church, but went out and built up congregations for himself." One such church was in Collins Co., one in Grayson Co., Tex.

Later Reverend Jarrell became paster of Walter Grove Church. This church became a revival center for the Cumberland Presbyterians in that region. When he left Walnut Grove, Brother Jarrell moved to Denton Co., Tex. Here he bought a farm and organized another church. He remained here 20 years. He died at Gainesville, Denton Co., Tex., November 7, 1909.

Sources: Jarrell Military Files, National Archives; Kentucky Adjutant General, *Report of the Adjutant General of Kentucky, 1861-1866,* 2 vols., Frankfort, Ky., 1866-1867, Vol. 1, p. 68, Vol. 2, p. 930; W.P. Kloster, "Rev. C.N. Jarrell," in *Our Senior Soldiers: The Biographies and Autobiographies of Eighty Cumberland Presbyterian Preachers,* Nashville, Tenn., 1915, pp. 81-83..

* * * * *

Adam Rankin Johnson

Adam Rankin Johnson served as a scout at Sacramento. After Shiloh, Forrest released his two scouts, Johnson and Robert Martin, to Maj. Gen. John C. Breckinridge. After completing an assignment into Kentucky, Johnson and Martin raised a regiment that became the Confederate 10th Kentucky Partisan Rangers but was later renamed 10th Kentucky Cavalry. Johnson served as colonel commanding; Martin as lieutenant colonel. After operating behind the lines in the area between the Green and Cumberland Rivers, the unit withdrew into Tennessee and joined John Hunt Morgan's cavalry division.

Adam Johnson became a brigade commander under Morgan. When Morgan was captured in Ohio during his 1863 raid into the North, Johnson managed to bring some of the Confederate troopers out of the trap and back South. In spite of attempts by Maj. Gen. Braxton Bragg to dismount and scatter Morgan's former soldiers, Johnson managed to keep them together as cavalry. After Morgan escaped from the Ohio Pentitentiary, Adam Johnson left Morgan, was promoted to brigadier general and placed in command of the Confederate District of Western Kentucky. In August 1864, during an engagement at Grubb's Cross Roads, near Princeton, Caldwell Co., he was wounded accidentally by one of his own men. The ball struck one eye, traveled across the bridge of his nose and took out the other eye.

After the war Johnson returned to Texas. He slowly regained his wealth. Using his remarkable memory of places he had seen, he became a real estate dealer and developer. In the 1880s he laid out, again from memory, and founded the town of Marble Falls, Burnet Co., on the Colorado River of Texas. "He selected the location," an 1899 article pointed out, " and the town had its conception in a brain that can only picture its streets and public buildings and its picturesque beauty in imagination."

"General Johnson can be seen upon the streets (of Marble Falls) almost every day during business hours," a correspondent of the *St. Louis Republic* wrote at the end of the 19th century. "He walks with his head erect, much faster than many younger men who possess all their faculties, and no one would ever dream that the man was pushing along in darkness were it not that he is usually accompanied by a little boy who

photo from Adam R. Johnson's Partisan Rangers of the Confederate Army

holds one of his hands. He seems to be perfectly familiar with the step and voice of every human being in the town, and seldom makes a mistake in calling the names of persons who pass or address him. He has recently been engaged in constructing a mill and harnessing water power to furnish the city with water. Everything has been done according to his plans and under his direct supervision. He closely examines every piece of machinery, every rock, and every wall." Adam Rankin Johnson died in 1922. The town of Marble Falls still exists.

Sources: Basil W. Duke, *History of Morgan's Cavalry,* (Cincinnati, 1867), contains material on the later career of Johnson; Albert Castel, *The Guerrilla War 1861-1865,* Special Issue, *Civil War Times Illustrated,* Oct. 1974, pp. 11, 30; "General Adam Rankin Johnson," *Owensboro* (Ky.) *Messenger,* Dec. 24, 1899; "Gen. Adam R. Johnson," *Confederate Veteran,* Vol. 8 (1900), pp. 117-118, p. 117 contains the description of him in Marble Falls; Adam R. Johnson, *The Partisan Rangers of the Confederate States Army,* William J. Davis, ed., 1904, rep. Evansville, Ind., 1971, contains items by numerous persons who knew Johnson and who give a review of his military career; Johnson File, National Archives; C. Brian Kelly, "Badly wounded by his own men during the Civil War, 'Stovepipe' Johnson still would pursue his lifelong dream, *Military History,* April 2002, p. 82; John H. Boyd, "Greenville Confederate almost captured town," *Times-Argus Messenger Magazine,* Central City, Ky., Sept. 18, 1969; J. Robert Smith, "Lt. Col. Robert Martin was talk of Gen. Forrest's army," *Times-Argus Messenger Magazine,* Central City, Ky., Sept. 11, 1969; Edmund L. Starling, *History of Henderson County, Kentucky,* Henderson, Ky., 1887, pp. 713-717; Ed Porter Thompson, *History of the Orphan Brigade,* 1898, rep. Dayton, O., 1973, also mentions Johnson's career; Ezra J. Warner, *Generals In Gray: Lives of the Confederate Commanders,* Baton Rouge, La., 1959, p. 156.

* * * * *

David Campbell Kelley

Maj. David C. Kelley, Forrest's deputy commander at Sacramento, became a lieutenant colonel when Forrest's regiment was organized. After Forrest became a general, Kelley took command of the regiment and was promoted to colonel. He later turned down a brigadier general commission, preferring to remain in command of his regiment. A graduate of Cumberland University, Lebanon, Tenn., Kelley had served as a missionary in China, 1854-1856. This made him unique in the Confederate Army; he was probably one of the few Southern officers who spoke fluent Chinese. When the war broke out he was pastor of a Methodist Church at New Market, Ala., not far from Huntsville. Here, in New

Market, he raised a cavalry company which called itself Kelley's Troopers. During the war, in addition to serving as deputy commander and later commander of the regiment and later still commander of a brigade, Kelley also served on Bedford Forrest's staff as unofficial chaplain and aide. During the war, he became known as "Forrest's fighting preacher."

After the war Kelley returned to preaching and also completed a doctorate of divinity at Cumberland in 1868. He served as pastor of several large Methodist churches in Tennessee, and was especially noted as a pastor at McKendree. One observer claimed in 1887: "What they do at McKendree Church, in Nashville, is felt to a greater or lesser extent to the borders of Southern Methodism."

Dr. David Kelley was involved in several important movements among Southern Methodists, including plans to establish a central university which finally resulted in Vanderbilt University. His name appears on the first decree which constituted the main portion of the charter of Vanderbilt University entered in the Minute Book of Chancery Court in Nashville on August 6, 1872. (The university at that time was named The Central University of the Methodist Episcopal Church South. On June 16, 1873, the name was changed, in a second decree, to Vanderbuilt Universtiy.) Doctor Kelley was associated with the university for 15 years as a trustee.

Kelley was also prominent in the plan to establish the Woman's Board of Missions and the Nashville College for Young Ladies. He worked hard to ensure the success of Sunday Schools within the Southern Methodist churches and was a leader in the Prohibitionist movement in Nashville. Doctor Kelley's son-in-law, Dr. Walter R. Lambuth, later elected bishop in the Methodist Church South, became "the outstanding missionary in the history of the Methodist Episcopal Church South," making an extensive tour into the African interior and helping to establish the mission work in the then Belgian Congo. Rev. Lambuth had married Daisy Kelley when he served as her father's assistant.

Dr. Kelley was also active in veterans organization, especially the Association of Lieut. Genl. N.B. Forrest Escort and Staff. He usually served as chaplain at the reunion meetings. He officiated at many weddings of his former soldiers' sons and daughters, or their other kin, and also officiated at many of his former soldiers' funerals.

Dr. David Campbell Kelley died in Nashville on May 15, 1909.

Sources: Adrian A. Bennett, *Missionary Journalist In China*, Athens, Ga., 1983, p. 20; Michael R. Bradley, *Nathan Bedford Forrest's Escort and Staff*, Gretna, La., 2006, pp. 40-41, 137-193 passim; Don H. Doyle, *Nashville in*

the New South: 1880-1930, Knoxville, Tenn., 1985, pp. 121-122, 134;
David Knapp Jr., *The Confederate Horsemen*, N.Y., 1966, pp. 244-247;
"Rev. D.C. Kelley," *Confederate Veteran*, Vol. 17 (1909), p. 421; *Tennesseans In The Civil War*, Nashville, Tenn., 1964, p. 55; *Tennessee: The Volunteer State: 1769-1923*, Nashville, 1923, Vol. II, p. 164; H.T. Tipps, "History
of McKendree United Methodist Church, in *Seven Early Churches of
Nashville*, Nashville, Tenn., 1972, pp. 7-21, pp. 15, 19, 21; "Charter"
Vanderbuilt University, January 4, 2006 <http://www.vanderbilt.edu/
boardoftrust/charter.html>, pp. 1-3.

* * * * *

Robert H. King

Union Lt. Robert King, who was slightly wounded during the
Sacramento fighting, received a promotion to captain in January 1862, to
fill the vacancy created by Capt. Albert G. Bacon's death. His promotion
to major came the following August, and he was made lieutenant colonel
in June 1863. In February 1864, King was appointed commander of the
3d Kentucky Cavalry. As Colonel King was being mustered out, the
officers and soldiers of the 3d Kentucky presented him with a beautiful
sword "as a testimonial of their high appreciation of him as a soldier and
gentleman." Although his National Archives files show him only as a
lieutenant colonel, the Kentucky Adjutant General's report shows King
receiving a brevet colonelcy at war's end.

Robert King had been and was after the war a printer in Frankfort,
"one of the very best of the craft." In early June 1866, King visited some
friends in the country and on the morning of Saturday, June 9, 1866
"complained of a slight indisposition." In spite of the slight illness, King
decided to walk back to Frankfort. He was found dead later in the day on
the roadside near the farm of Capt. H.I. Todd. It was believed his death
probably resulted from "overheating himself while laboring under
temporary indisposition." He was buried with military honors, with "a
large concourse of citizens," on the following Sunday.

Sources: "The Affair at Sacramento," *Frankfort* (Ky.) *Yeoman*, Jan. 2, 1862;
      Kentucky Adjutant General, *Report of the Adjutant General of Kentucky,
      1861-1866*, 2 vols., Frankfort, Ky., 1866-1867, Vol. 1, pp. 63, 67, Vol. 2, p.
      931; King File, National Archives; "Col. Robert H. King," *Tri-Weekly
      Yeoman*, Frankfort, Ky., June 12, 1866.

* * * * *

Robert Maxwell Martin

Robert M. Martin, one of Forrest's scouts at Sacramento, joined with
fellow scout Adam Johnson to raise the Confederate 10th Kentucky

Partisan Rangers, later 10th Kentucky Cavalry. When Johnson became a brigade commander, Robert Martin took command of the 10th. In June 1864, he was wounded at Mt. Sterling, Ky. The wound disabled him for cavalry duty so Martin was sent to Canada by Confederate Secretary of State J.P. Benjamin. From Canada Martin harassed the North. Among his activities was the attempt to burn New York City in 1864.

Following the war, Martin engaged in the tobacco business in Evansville, Ind., Louisville, Ky., and occasionally in New York. In Evansville, he operated a tobacco business at 1st and Ingle Streets, in company with Lee M. Gardiner. He lived in a house at the corner of 1st and Locust Streets. From there he moved to Louisville to engage in the wholesale tobacco business and later moved to New York. He reportedly made and lost more than one fortune.

In the 1870s, Martin married Caroline Wardlaw, daughter of a South Carolina Supreme Court justice. In January 1901 Robert Martin died under mysterious circumstances in New York City. Suspicion lies strongly on his wife killing him, although officially he died from the continuing effects of his wartime wounds. At the very least, it appears his wife allowed him to die in a dingy apartment in New York. She claimed to be broke, forgetting to mention a life insurance policy. Confederate veterans in the New York area took up a collection to bury Robert Martin honorably. Mrs. Martin never bothered to repay them.

Caroline Martin was later convicted of manslaughter in the death of her and Martin's daughter, although she was believed to have murdered the young lady. Her sister, whose testimony would probably have convicted Caroline, died before the trial was held. Caroline was actually believed to have been insane earlier and did die in a hospital for the criminally insane in 1913. She was never charged in Robert's death.

Sources: Basil W. Duke, *History of Morgan's Cavalry,* (Cincinnati, 1867), contains material on the later career of both Johnson and Martin; Nat Brandt, *The Man Who Tried To Burn New York,* Syracuse, N.Y., 1986; Albert Castel, *The Guerrilla War 1861-1865,* Special Issue, *Civil War Times Illustrated,* Oct. 1974, p. 11; Alvin F. Harlow, *Murders Not Quite Solved,* N.Y., 1938, pp. 165-207; John W. Headley, *Confederate Operations in Canada and New York,* (N.Y., 1906); Adam R. Johnson, *The Partisan Rangers of the Confederate States Army,* William J. Davis, ed., 1904, rep. Evansville, Ind., 1971, contains numerous mentions of Martin; Oscar A. Kinchen, *Confederate Operations in Canada and the North,* North Quincy, Mass., 1970; Martin File, National Archives; Richard T. Martin, "Recollections of the Civil War," Otto A. Rothert, *A History of Muhlenberg County,* Louisville, 1913, pp. 285-317, pp. 301-306; Martin's biography in Otto A.

Rothert, *A History of Muhlenberg County,* Louisville, Ky., 1913, pp. 318-325; John H. Boyd, "Greenville Confederate almost captured town," *Times-Argus Messenger Magazine,* Central City, Ky., Sept. 18, 1969; J. Robert Smith, "Lt. Col. Robert Martin was talk of Gen. Forrest's army," *Times-Argus Messenger Magazine,* Central City, Ky., Sept. 11, 1969; "Col. R.M. Martin became wealthy following war," *Times-Argus Messenger Magazine,* Central City, Ky., Sept. 18, 1969.

* * * * *

Calvin A. McCullough

Wounded Union Pvt. Calvin A. McCulloughh, Company A, 3d Kentucky, had been left at a farm house near where he collapsed. A local doctor, unnamed (called "a practitioner of the neighborhood"), gave initial treatment and administered purgatives. He also apparently looked in on the trooper with some regularity. Union Army surgeons also occasionally checked on McCullough but their visits were limited because the farm house was located "between the lines," a gray area just south and west of Green River controlled by neither side. "After protracted suffering," over an unstated period of time, trooper McCullough recovered enough to return to duty and "performed some little service."

While the occupants of the farm house are not named, there is a chance this was the farm owned by Jessie Plain. According to a tradition in the Stroud family, Kate Plain, as an 18-year-old, was spending time with her brother, Jessie. One day while she was there, two wounded Union soldiers came to the farmhouse for help. They were put to bed and tended but one soldier died. Later the second soldier was taken away by members of his troop. Kate Plain later married and was the mother of Ernest Stroud, who told the story as related by his mother.

Private McCullough appears to have returned to duty with the 3d Kentucky Cavalry during March 1862 and was promoted to quartermaster sergeant of his battalion on March 30. He, in fact, appears to have been assigned as company quartermaster sergeant by the end of February 1862. However, his wounds continued to bother him and on September 18, 1862, he was discharged on a certificate of disability. By that time he was serving as a commissary sergeant. According to the certificate, Sergeant McCullough had been unfit for duty for at least the prior 60 days. Regimental Surgeon William Singleton reported that after his discharge McCullough "enjoyed better health than could be expected." His final residence and date of death are unknown.

Sources: Kentucky Adjutant General, *Report of the Adjutant General of Kentucky, 1861-1866,* 2 vols., Frankfort, Ky., 1866-1867, Vol. 1, p. 64; Military File, National Archives; *Medical and Surgical History of the War of the*

Rebellion, Joseph K. Barnes, ed., 6 vols., Washington, D.C., 1870-1888, Vol. 2, pt. 2, p. 76 (listed as Dec. 1862); Thomas Southard, "Sacramento sixth graders report more local Civil War stories," *McLean Co. News* (Calhoun, Ky.), Nov. 27, 1958.

* * * * *

William Sugars McLemore

Capt. William Sugars McLemore, whose detachment of the 8th Tennessee Battalion joined Forrest for the late December scout, was a graduate of Transylvania University, Lexington, Ky., and Lebanon Law School, in Wilson County, Tenn., He had been pre-war Williamson County, Tenn., Court Clerk. The 8th Battalion served as the core for the 4th Tennessee Cavalry Regiment (first known as 3rd Tennessee Cavalry Regiment), created on May 26, 1862. McLemore is shown in some reports with the rank of major in the new regiment but in *That Devil Forrest,* John Allan Wyeth reports that Captain McLemore led the 4th Tennessee when Starnes first commanded Forrest's brigade. In March 1863 McLemore was named acting commander of the 4th Tennessee and was promoted to major June 1, 1863. On Feb. 25, 1864 he was promoted to colonel and appointed an acting bragadier in November 1864. In the last month of the war he commanded a brigade. He led his brigade, Dibrell's Tennessee Brigade, as part of the escort of Confederate Pres. Jefferson Davis as the latter attempted to cross the Mississippi River and join the Confederates still fighting in the TransMississippi sector. They didn't make it. William S. McLemore surrendered on May 9, 1865.

Following the war, in 1878, McLemore was elected Circuit Judge, Ninth Tennessee Circuit, a position he held for 14 years. After leaving the bench he moved from Franklin to Murfreesboro and practiced law in the firm of McLemore and Richardson. Just prior to 1900 he suffered a slight stroke. He died at Murfreesboro, Tenn., in August 1908.

Sources: *Confederate Patriot Index: 1924-1978,* Vol. II, Columbia, Tenn., 1978, p. 363; Michael Cotten, *The Williamson County Cavalry: A History of Company F, Fourth Tennessee Cavalry Regiment, C.S.A.,* n.p., 1994, pp. 2-3, 214; Michael Cotten, *Williamson County Confederates,* n.p., 1996, p. 221; Joseph R. Haw, "The Last of C.S. Ordnance Department," *Confederate Veteran,* Vol. 22 (1914), pp. 450-452, p. 451; Robert Selph Henry, *"First With the Most" Forrest,* 1944, rep. N.Y., 1991, pp. 145, 156-157, 193, 201, 351; B.L. Ridley, "Chat With Col. W.S. McLemore," *Confederate Veteran,* Vol. 8 (1900), p. 262; Bromfield L Ridley., "Chat With Col. W.S. McLemore," *Battles and Sketches of the Army of Tennessee,* Mexico, Mo., 1906, pp. 177-181; Stewart Sifakis, *Compendium of the Confederate Armies: Tennessee,* N.Y., 1992, p. 45; *Tennesseans In The Civil War,*

Nashville, Tenn., 1964, pp. 62-64; John Allan Wyeth, *That Devil Forrest,*
1899, paperback rep. Baton Rouge, La., 1989, pp.28, 147, 176.

* * * * *

Charles Edward 'Ned' Meriwether

"Had he lived," Col. Forrest said after the battle, "Capt. Ned
Meriwether would have been one of the great cavalry officers of the
Confederacy." Capt. Joseph M. Williams took command of the company
Capt. Charles Edward 'Ned' Meriwether had been recruiting. Captain
Williams kept the company with Colonel Forrest until after they escaped
from Fort Donelson. The company then joined Col. Benjamin Hardin
Helm's 1st Kentucky Cavalry.

Lt. J.B. Wheeler took Captain Meriwether's body home to Todd Co.
There Meriwether was buried in the family's cemetery. He would be
posthumously promoted by his civilian kin to the rank of colonel. Eleven
other Confederate dead would join him in the cemetery, which is main-
tained by the Meriwether family. But the story did not end there.

Caroline Meriwether Sturdivant was Ned Meriwether's younger
sister. Her marriage to John Sturdivant had ended in divorce. Now she
had lost a favorite brother. Throwing herself into the cause of the South,
her rooms and, when necessary, her barn became meeting places for like-
minded women. Here they met to sew, knit and make bandages and
clothing for the soldiers. Caroline, an excellent horsewoman—her
father's horses were famous throughout the country during pre-war
years—sometimes would carry medicine and clothing through the lines.

When the war ended, Caroline's father's fortune was largely gone
and Caroline had a son to raise. She sold her share of the Meriwether
estate and moved to Nashville so her son could receive a better education. In Nashville Caroline met Col. Michael Campbell Goodlett, a
lawyer and widower with four children. They married in 1869. In 1871 a
daughter was born to them. A few years later Caroline's father died,
followed in death by her son, as he was finishing at Vanderbilt. Her
daughter and son-in-law died on the same day in a typhoid epidemic.
They left Caroline with two small boys to raise. Colonel Goodlett died of
typhoid fever three weeks later.

In the meantime Caroline had become very active in organizations
aiding Confederate veterans as well as associations and committees
honoring fallen Confederates. From these came an idea. Her dream as
well as the dream of other Southern women eventually saw fruit with the
formation of the United Daughters of the Confederacy (UDC) which
largely grew from Caroline's push. She served as the first president,
elected September 10, 1894.

Sources: J.H. Battle, "Todd County History," in J.H. Battle and W.H. Perrin, eds., *Counties of Todd and Christian, Kentucky,* Chicago and Louisville: 1884, p.110; Elizabeth Cody Ellis Bachman, "Our Founding," *United Daughters of the Confederacy Magazine,* Sept. 1969, p. 1; Edna S. Macon, "Legacy of the Skirmish at Sacramento: the United Daughters of the Confederacy," *Kentucky's Civil War 1860-1865,* 2002-2003 edition, Clay City, Ky., 2002, p. 14; Louisa H.A. Minor, *The Meriwethers and Their Connections,* Albany, N.Y., 1892, pp. 36, 132; William Henry Perrin, ed., *Counties of Christian and Trigg,* Louisville, Ky., 1884, p. 176; Ed Porter Thompson, *History of the Orphan Brigade,* 1898, rep. Dayton, O., 1973, pp. 881, 1009; Ed Porter Thompson, *History of the Orphan Brigade,* 1898, rep. Dayton, O., 1973, p. 881; Josephine Turner, "Caroline Meriwether Goodlett: Founder, United Daughters of the Confederacy," *United Daughters of the Confederacy Magazine,* Sept. 1962, pp. 12-14, 36-41, p. 36; Josephine M. Turner, *The Courgeous Caroline Founder of the UDC,* Montgomery, Ala., 1965, pp. 27, 59; Frances Marion Williams, *The Story of Todd County, Kentucky, 1820-1970,* Nashville, 1972. pp. 81. 397-398.

* * * * *

Mary Susan 'Mollie' and Sarah E. 'Bettie' Morehead

As to the young lady who aided the Confederates at Sacramento, Mary Susan 'Mollie' Morehead survived the Civil War by only five years. Only 18 years old when she rode to report to Colonel Forrest, Mollie Morehead (on official records, her family name is spelled both Morehead and Moorehead) was born March 16, 1843. She was the daughter of Hugh N. and S. Elizabeth Morehead. Her sister, Sarah E. 'Bettie' Morehead was born March 1, 1845, and thus was only 16 when she left her sister to ride and warn her father and brothers that Yankee cavalry were in the vicinity.

On May 16, 1860, Daniel and Catherine Whitmer had deeded Hugh N. Moorehead (sic) of Butler Co., Ky., 158 acres of land on Cypress Creek. Hugh was pro-Southern and a slave owner; thus, he needed to know that Federal forces were nearby.

There is no other infomation about the wartime activities and events surrounding the Morehead family. After the war, on November 14, 1866, Mollie married Dr. George E. Stowers, a dentist. She died just over three years later, on March 29, 1870, apparently during childbirth. An infant son is listed as born and died the same day.

Bettie married Daniel K. Nall September 4, 1867. She survived until November 20, 1914.

Sources: Elizabeth S. Cox, correspondence, July 20, 1971; Elizabeth S. Cox, correspondence, undated, quoting McLean Co. Deed Book B, p. 470; Mrs. J.C. Hill, copier, "Cumberland Presbyterian Church Cemetery," in *McLean County, Kentucky Cemeteries, Volume I,* Owensboro, Ky., 1977, pp. 67-76, p. 68; McLean Co., Ky., Will Book I, under date March 20, 1875; McLean Co., Ky., Marriage Book A, p. 56 (Mollie S. Moorehead [sic] to Dr. George E. Stowers) and p. 662 (Sarah E. Moorehead [sic] to Daniel K. Nall); "Girl on the Firing Line," *McLean Co. News,* Dec. 28, 1961; Gina Hancock, "Anniversary of battle is this week," quoting, among others, Cora Bennett, Elizabeth Cox, Martha Harris, Allen Taylor Nall and Kenny Ward, *McLean Co. News,* Dec. 31, 1987; "NEWS reader identifies heroine of Sacramento battle," quoting Cora Bennett, *McLean Co. News,* Feb. 22, 1962.

* * * * *

Eli Huston Murray

Eli H. Murray, the 18-year-old Union commander at Sacramento, started his rise with the fight there. General Crittenden wrote of Sacramento that "by the testimony of all, Major Murray's conduct in the field deserves the highest praise." When Col. James S. Jackson, 3d Kentucky Cavalry commander, was placed in command of the cavalry brigade after Shiloh, Murray was placed in command of the 3d Kentucky. Shortly after Jackson was promoted to brigadier general, Murray, on August 13, 1862, was promoted to colonel. He was 19 years old.

photo from *Biographical Encyclopaedia of Kentucky...*

Throughout the remainder of 1862 the 3d Kentucky was engaged in constant operations, including frequest skirmishing with Southern cavalry. With the movement from Nashville toward Murfreesborogh, Tenn., the 3d Kentucky Cavalry operated on the extreme left of the Federal force. Murray again earned a mention for his gallantry. At Laverne, Tenn., on December 26, 1862, he led his command in support of infantry who were being attacked.

The Battle of Stone's River opened at dawn, December 31, 1862, when the Confederates struck the Federal right and began rolling it up. In some areas the Federals broke and ran. In one area, Colonel Murray counterattacked, charging Southern cavalry. Col. R.H.G. Minty, Murray's brigade commander, wrote that Murray, "with a handful of men, performed services that would do honor to a full regiment." Murray had with him only about 80 men.

From February 1863 until December 1863, the 3d operated in Kentucky with headquarters at Hopkinsville. In December Colonel Murray and the 3d Kentucky returned to General Rosecrans' Army of the Cumberland. In March 1864 the unit was veteranized. At the end of a 30-day veteran's furlough, the men reported for duty at Chattanooga, Tenn. where they joined the 3d Cavalry Division, part of General Sherman's drive toward Atlanta, Ga. Murray wrote at one point that it was "impossible for any one not a participant to have a conception of the many marches made and successful engagments."

On May 18, 1864, the young officer was temporarily placed in command of the 3d Cavalry Division. General Sherman is quoted as later saying, "Whatever I told Murray to do, I knew it would be done." High praise for a man who became a division commander at the age of 21. On March 25, 1865, at the age of 22 years, one month and 15 days, Murray was brevetted brigadier general of volunteers. Four months later, on July 15, 1865, General Murray was discharged along with the rest of his old 3d Kentucky Cavalry Regiment. In May 1864 then Colonel Murray had professed that he was a proud man, "in the first place because I am alive; in the next place that I am a Kentuckian..."

Following the war, Murray practiced law, served as a U.S. marshal and manager of the *Louisville* (Ky.) *Commercial*. In 1880 Eli was appointed territorial governor of Utah and reappointed in 1884, but resigned in 1886. The Murrays then moved to San Diego, Calif., where Eli edited the *California American* and was involved in land and railroad deals. In 1896, in ill health, he returned to Kentucky. At 3:30 p.m., November 18, 1896, at the home of his father-in-law in Bowling Green, Eli Murray died of diabetes. (An interesting note here is that both the Union and Confederate commander at the Battle of Sacramento probably died of diabetes.)

Sources: *Biographical Encyclopaedia of Kentucky of the Dead and Living Men of the Nineteenth Century* (Cincinnati, 1878), p. 275; J. D. Cox, *March To The Sea*, 1883, rep. N.Y., 1963., pp. 24, 32 et seq.; H.H. Crittenden, comp., *The Crittenden Memoirs*, N.Y., 1936, pp. 528-529; Crittenden's report, *War of the Rebellion: A Compilation of the Official Records of the Union and Confederate Armies* (commonly *Official Records of the Union and Confederate Armies*), 128 vols., index, and atlases, Washington, D.C., 1880-1901, Ser. I, Vol. 7, p. 63; David Evans, *Sherman's Horsemen: Union Cavalry Operations in the Atlanta Campaign*, Bloomington, Ind., 1996, scattered references; "Eli Murray Is Dead," *Breckinridge News*, Cloverport, Ky., Nov. 25, 1896; "Eli H. Murray Dead," *Louisville* (Ky.) *Courier-Journal*, Nov. 19, 1896; "Gen. Eli Murray," *Messenger*, Owensboro, Ky.,

Nov. 19, 1896; Thomas Marshall Green, *Historic Families of Kentucky* (Baltimore, 1964), p. 250; James Moore, *Kilpatrick and Our Cavalry*, N.Y., 1865, pp. 165-195; Murray Military and Pension Files, National Archives; Stewart Sifakis, *Who Was Who in the Union*, N.Y., 1988, p. 284; William B. Sipes, *The Seventh Pennsylvania Veteran Volunteer Cavalry*, Pottsville, Pa., n.d., pp. 123-135; Thomas Speed, *The Union Cause in Kentucky, 1860-1865*, N.Y., 1907, pp. 69-70; Richard G. Stone Jr., *Kentucky Fighting Men 1861-1945*, Lexington, Ky., 1982, p. 11; Union Soldiers and Sailors Monument Association, *The Union Regiments of Kentucky*, Louisville, 1897, pp. 69-70.

* * * * *

James Wilburn Starnes

Lieutenant Colonel James Wilburn Starnes led the 40-man detachment from his 8th Tennessee Cavalry Battalion at Sacramento. He was "subsequently distinguished as one of Forrest's colonels." In May 1862 the 8th served as the core for the 4th Tennessee Cavalry Regiment, originally called the 3d Cavalry Regiment, with Starnes as colonel commanding. By late March 1863 Colonel Starnes was commanding "Forrest's Old Brigade," 2d Brigade of Forrest's Division. By late-June 1863 Starnes had been nominated for promotion to brigadier general, although the promotion had not yet come through.

Beginning June 23, 1863, and continuing for the next week, the series of skirmishes and maneuvers known as the Tullahoma Campaign forced Confederate Gen. Braxton Bragg to withdraw behind the Tennessee River. Bragg had badly weakened his cavalry forces. First he sent part of his mounted forces to Mississippi in an attempt to aid General Pemberton at Vicksburg. He next gave in to Brig. Gen. John H. Morgan's repeated requests to stage another raid into Kentucky. Morgan pulled more than 2,000 troopers away on what turned into his disastrous raid into Ohio. Bragg was left with only about 9,000 cavalry to cover the Confederate flanks, screen the front and scout Union General Rosecrans advancing Federal force, which included more than 12,000 mounted troopers. Not nearly enough.

By Sunday, June 28, Bragg's Confederate cavalry had been used, abused, mauled and worn out. Still they attempted to halt the Union advance. Shortly after 2 p.m. that day, a strong advance force from the 3d Infantry Brigade, 3d Division, Union XIV Corps encountered Colonel Starnes' small brigade. Colonel Starnes rode forward to encourage his men while he searched for a good spot to place an artillery battery. Several of his men requested he move back out of rifle range. He thanked them for their consideration but remained near the front.

Colonel Starnes was accompanied by Capt. Daniel F. Wade, Company C, 3d Tennessee Infantry. Captain Wade, Starnes' pre-war neighbor, was still recuperating from a severe wound received at Ft. Donelson and had preceded his unit then being transferred to Bragg. He, too, suggested Yankee sniper fire was getting awfully close and maybe they should withdraw. Again Starnes declined to withdraw but suggested the captain should, if he felt threatened. Like Forrest, Starnes tended to be brave to a point of recklessness. Wade was also brave and he remained.

At around 3 p.m. Colonel Starnes paused under a large tree, apparently studying the situation. By now, Federals had started to move around the flanks of the Confederate cavalry. Suddenly, a sharpshooter's aim proved true. The bullet struck Starnes in the side at the waist, passed through a kidney and came out the front on the other side.

Captain Wade, slightly behind Colonel Starnes, realized the colonel had been shot. He rode forward and grabbed both his own reins and Starnes' in his left hand. He held on to the colonel with his right arm. The captain managed to make it to the picket lines of the Confederate 48th Tennessee Infantry before the pain in his unhealed right hip caused him to faint. Both he and Starnes fell to the ground. The colonel was dead.

Pvt. J.M. Jackson, Company I, 4th Tennessee Cavalry, claimed to have happened to see the smoke puff from the Federal sniper's rifle when he fired the shot that killed Colonel Starnes. Keeping trees between himself and the sharpshooter, Jackson made his way to within about 50 yards of the shooter. The Confederate then shot the Federal from his perch. Jackson then grabbed the sniper's Whitworth rifle and took it with him. The private had his short spell of fame for avenging Colonel Starnes' death.

The Battle of Sacramento saw the beginnings of a personal friendship between Forrest and Starnes. Their mutual respect lasted until Starnes' death. This friendship and respect came in spite of the great differences in their personalities and command styles. Colonel Starnes was smaller than average and well educated while Forrest was large in size with only a limited education. The colonel politely and quietly gave orders, often requesting subordinate to perform some order during the heat of battle; Forrest shouted obscenities and yelled at commanders on the field. As Pvt. J.R. Harris summed up the difference: "He (Colonel Starnes) was a kind hearted man, and could lead brave men farther than most men, while Forrest could make a coward fight."

Sources: Mark M. Boatner III, *The Civil War Dictionary,* 4th printing, N.Y., 1966, "Tullahoma Campaign"; Michael Cotten, *The Williamson County*

Cavalry: A History of Company F, Fourth Tennessee Cavalry Regiment,
C.S.A., n.p., 1994; Michael Cotten, *Williamson County Confederates,* n.p.,
1996, p. 227; Robert Selph Henry, *"First With the Most" Forrest,* India-
napolis and N.Y., 1944, rep. N.Y., 1991, pp. 145, 156-157, 193, 201, 351;
Thomas Jordan and J.P. Pryor, *The Campaigns of Lieut.-Gen. N.B. Forrest,*
and of Forrest's Cavalry, 1868, rep. Dayton, Ohio, 1973, footnote, p. 49;
Stewart Sifakis, *Compendium of the Confederate Armies: Tennessee,* N.Y.,
1992, p. 45; H. Gerald Starnes, *Forrest's Forgotten Horse Brigadier,* Bowie,
Md., 1995, Private Harris quote from page 89; *Tennesseans In The Civil*
War, Nashville, Tenn., 1964, pp. 62-64; John Allan Wyeth, *That Devil*
Forrest, 1899, rep. Baton Rouge, La., 1989, pp. 28, 147, 176.

* * * * *

John L. Walters

2d Lt. John L. Walters, Company B, 3d Kentucky Cavalry, captured
at Sacramento, was exchanged for 2d Lt. E.M. Ware, 5th Virginia
Cavalry, on September 21, 1862. While a prisoner, on January 10, 1862,
Walters had been promoted to first lieutenant. Lieutenant Walters
returned to duty on October 7, 1862. But his military career ended under
a cloud.

On June 21, 1863, Lieutenant Walters was placed under arrest. At a
General Court Martial held at Hopkinsville, Ky., on July 24, 1863, 1st Lt.
John L. Walters was arraigned and tried on three separate charges:

1st Charge: "Leaving Camp without permission." The specification
states that he left camp without permission "and did remain thus about
twenty four hours, all this at Hopkinsville, Ky., on or about the 17th day
of June 1863."

2d Charge: "Resisting the officer of the Guard, while in performance
of his duty." The specification here states the lieutenant, "did, while
under arrest, on march from Hopkinsville, Ky. to Clarksville, Tenn. resist
and threaten the life of Lieut. (Edward) Kelly, 3d Ky. Cavalry who had
been regularly detailed, and was acting as officer of the guard,... on or
about the 10th day of July, 1863."

3d Charge: "Conduct unbecoming an officer and a Gentleman." The
specification states Lieutenant Walters "did strike and threaten to shoot
Corporal George A. Boarman, 3d Ky. Cavalry, all this at Russellville, Ky.
July 13th, 1863."

Lieutenant Walters pled not guilty to the first charge but guilty to the
second and third charges, "except so much as says 'did threaten to
shoot'."

The court, "after Mature deliberation on the evidence adduced,"
found 1st Lt. John L. Walters guilty on each charge and its specification.
The court sentenced Lieutenant Walters, "To be cashiered from the

service of the United States." Passed forward to Headquarters, Department of the Ohio, at Cincinnati, the "proceedings, findings and sentence" in the case were "approved and confirmed" and Lieutenant Walters ceased to be an officer "in the Military Service of the United States." Officially, he was dismissed by order of the President, February 14, 1864.

In the end, however, John Walters may have redeemed himself. On August 27, 1864, Capt. Jake Bennett, Confederate 10th Kentucky Cavalry, with 19 men, charged into Owensboro. The Southerners forced everyone found on the street, about 300, to go to the courthouse. The provost marshal and some other govenment officals fled. The Confederates killed three members, 108th U.S. Colored Infantry, guarding Union govenment stores at Ayer's Wharfboat, then burned the wharfboat. The Confederates set fire to govenment stores on the levee just before they left, but citizens saved most of this property after the raiders withdrew. The raiders destroyed freight and private property worth at least $6,000. The only resistance they met came from former lieutenant John L. Walters. He attempted to shoot it out with the Confederates and was shot and killed.

Sources: Aloma Williams Dew, "'Between the Hawk and the Buzzard': Owensboro During the Civil War," *The Register of the Kentucky Historical Society,* Vol. 77, No. 1 (Winter 1979), pp. 1-14, pp. 8-9; *History of Daviess County, Kentucky,*1883, rep. Evansville, Ind., 1966, pp. 172-173; Keith Lawrence, "Confederate day all but forgotten," Owensboro, Ky., *Messenger-Inquirer,* June 3, 1977; *3d Ky. Cav. Memo. Book,* National Archives, pages unnumbered; Arthur C. Truman Jr., *At War With Ourselves 1861-1865,* Utica, Ky., 1992, p. 26 (dates as Aug. 3); General Court Martial Orders, Co. B, 3d Kentucky Cavalry, Muster Out Rolls,Walters Military File, National Archives; "War Reminiscences," *Owensboro* (Ky.) *Messenger,* Jan. 29, 1888; *The War of the Rebellion: A Compilation of the Official Records of the Union and Confederate Armies,* Washington, D.C., 1880-1901. Series II, Vol. 4, , p. 581.

* * * * *

John L. Williams

Pvt. John L. Williams, Company D, Union 3d Kentucky Cavalry, fought gallantly at Sacramento, even throwing his empty revolver at Colonel Forrest and nearly knocking the Southerner off his horse. Forrest admired courage and quickly moved in to stop his men after they attacked the unarmed Federal cavalryman. Later, rather than taking the injured Williams south as a prisoner-of-war, Forrest paroled the Union private and left him in the care of his farmily at Greenville.

Later, passing through Greenville, Forrest stopped and checked on the Union private. Williams' wife and children "displayed so much genuine feeling and gratitude" for Forrest's actions that, as the latter left the house, "he was seen to wipe a visible tear from his eye."

While not mortal Williams' wounds were serious. On October 8, 1862, at Louisville, Ky., John L. Williams was discharged on a Certificate of Disability. The certificate states that Williams was wounded on the right wrist, "severing the tendons on the palm... and completely disabling the right hand. Also two saber wounds of the skull, one of which penetrated the cranium, and has left a suture which is still unhealed." He also complained of "symptoms of compression of the brain." The Battle of Sacramento casualty was discharged as three-quarters disabled. Prior to the war John Williams had been a mechanic but from the time of his wounding until October 1862 he had been unable to work Or, as Williams and two witnesses had sworn, his occupation had been "nothing."

After the war Pvt. John L. Williams, in a long-standing American tradition, received an unofficial promotion to become "Captain" Williams. An article in early June 1887 states: "Capt. John L. Williams, an old soldier of the Third Kentucky Cavlary, who measured sabres with Gen. Forrest at the battle of Sacramento, and who bears upon his body the marks of Gen. Forrest's weapon, is now very ill and not expected to live. Gen. Forrest has long ago slept with those gone before, carrying upon his body the marks of the old Captain's blade." Of course, a few liberties were taken with the facts. It was not Forrest's saber that left it marks on Williams body nor did Forrest's body carrying the marks of the "old Captain's blade." However, Williams had fought Forrest man-to-man. The article had been right in reporting that Williams was not expected to live. At 8 p.m., Saturday, June 11, 1887, he died in Greenville.

Sources: "An Old Soldier Dying," *Owensboro* (Ky.) *Messenger,* June 10, 1887, quoting the *Central City* (Ky.) *Republican*; Thomas Jordan and J.P. Pryor, *The Campaigns of Lieut.-Gen. N.B. Forrest, and of Forrest's Cavalry,* 1868, rep. Dayton, Ohio, 1973, p. 55; "Local Laconics," *Owensboro* (Ky.) *Messenger,* June 14, 1887; Williams Military, Pension Files, National Archives

The Weapons

Colt Navy Revolvers

In early 1836, 22-year-old Samuel Colt patented his revolver. The U.S. military's disregard for the weapon resulted in bankruptcy for Colt's first company. But the Mexican War created a demand for Colt revolvers. The first type Colt manufactured in quantity for the government was the Model 1848, the 'Dragoon' or 'Holster Pistol.' The Dragoon was a huge, heavy four-pound, two-ounce, .44 caliber rifled pistol.

Three years later Colt produced the lighter 'Navy' model revolver. The six-shot, .36 caliber Navy was 13 inches long, with a 7 1/2 inch octagonal barrel rifled with seven grooves and a muzzle velocity of 760 feet per second. It weighed only two pounds, ten ounces. This revolver was designed to be worn in a holster carried on a person, something the heavy Dragoon largely precluded. The Navy quickly gained a reputation for reliability and by war's end at least 215,000 had been produced and sold. Especially popular in the Confederacy, the Navy served as the prototype for almost all Southern-made revolvers. (The 1861 model Colt Navy had a round barrel.)

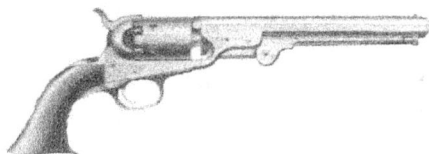

(In 1860 Colt patented a .44-caliber revolver which became the official model for the U.S. Army. Southerners, however, remained true to the Navy. Despite the latter's comparatively small caliber, the Navy and Southern-made copies were the main handguns used by the Confederate cavalry.)

Another asset of the Colt Navy was its versatility when it came to ammunition, since it could accommodate a wide variety of cartridge types. "Cartridge," of course, doesn't mean the common brass item used in modern weapons. Instead, the "cartridge" was a pre-packaged unit of black powder, wad and lead ball wrapped in paper, oil silk or animal gut ready to be inserted into the front end of the revolver's chamber. Provided the caliber was correct, one type was as good as another. So, the Navy's owner seldom had a supply problem. If by chance no prepared cartridges were available, the revolver could be loaded with loose powder and a ball rammed into the chamber, protected there by a smear of grease or fat. The only possible difficulty might arise with the supply of percussion caps. Once the chambers were loaded, a percussion cap was placed on the nipple of each chamber and the revolver was ready for action. Most Navy owners carried at least one extra cylinder loaded and capped. Exchanging cylinders was much quicker than loading individual chambers with powder and ball.

At close range the .36 Navy Colt , with high velocity and an almost flat ended round ball carried considerable knock-down shock when striking a human target. This attribute resounded well with fighters.

At the war's beginning, some Union cavalrymen carried cumbersome horse pistols, but they were soon replaced by revolvers. These revolvers included Remington and Savage although most were Colts.

Sources: William A. Albaugh III, and Edward N. Simmons, *Confederate Arms,* Harrisburg, Pa., 1957, pp. 7-9; Mark M. Boatner III, *The Civil War Dictionary,* 4th printing, N.Y., 1966; John S. Bowman, exec. ed., *The Civil War Almanac,* N.Y., 1983, pp., 282-283; Jack Coggins, *Arms & Equipment of the Civil War,* N.Y., 1962, p. 41; Elmer Keith, *Six Guns,* N.Y., 1961, pp. 13-14, 32; Claud E. Fuller and Richard D. Steuart, *Firearms of the Confederacy,* Huntington, W. Va., 1944, p. 244; Arcadi Gluckman, *United States Martial Pistols and Revolvers,* Buffalo, N.Y., 1944, pp. 174-185; Francis A. Lord, *Civil War Collector's Encyclopedia,* N.Y., 1965, pp. 206, 251; James C. Rikhoff, "Guns of the Civil War: Part IV: Confederate Revolvers," *Gunsport,* Sept. 1960, pp., 34-36, 73-76, p. 36; Bell I. Wiley, *Life of Billy Yank,* 1952, rep. Baton Rouge, La., 1978, p. 63.

Enfield Rifle Musket

The British Army adopted the Enfield Rifle Musket in 1855 and it remained their general infantry weapon until the adoption of a breech-loader in 1867. The standard Enfield weighed 8 pounds, 14 1/2 ounces, was 54 inches long with a barrel length of 39 inches. The Enfield fired a .577 caliber bullet. But the difference in size between that and the .58-caliber was so slight the bullets were basically interchangeable. The Enfield was considered very accurate up to 800 yards and fairly accurate at 1,100 yards. However, some considered the Enfield's accuracy beyond 700 yards questionable.

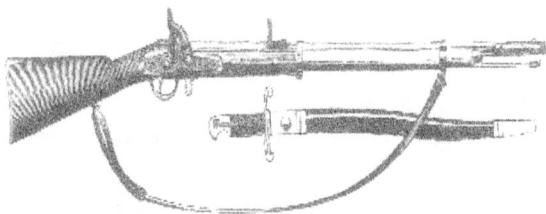

The Enfield was slightly lighter than the American Springfield; but, according to contemporary accounts, was a "beautiful arm and presented a natty appearance." One version, a .58-caliber rifled Enfield marked "LAC 1862" (for London Armory Co.) was exclusively supplied under contract to the Confederacy.

Many histories suggest that only the Confederacy used Enfields. In fact, the U.S. Government purchased 428,000 Enfields in the early months of the war, before Northern weapons production could meet the growing Union Army's needs. And the weapon was popular with Federals. As noted, some Federals at Calhoun, including some members of the 3d Kentucky Cavalry, had been issued Enfields. But the Confederacy imported many more over the course of the war. In fact, they imported 400,000 during 1861-1862 alone. And Wiley reports that the Enfield was "perhaps the most popular gun in the Confederate service" after 1862, and "one of the most effective."

Although Bedford Forrest had early shown a liking for shotguns and Maynard carbines for his men, he later came to prefer the greater power and reach of the muzzle-loading Enfield.

Sources: Wayne Austerman, "Maynard," *Civil War Times Illustrated,* April 1986, pp. 42-45, p. 43; Mark M. Boatner III, *The Civil War Dictionary,* 4th printing, N.Y., 1966; Ashley Halsey Jr., *Who Fired The First Shot?,* N.Y., 1963, "What They Fought With," pp. 48-81, p. 51; Francis A. Lord, *Civil War Collector's Encyclopedia,* N.Y., 1965, pp. 240, 247; Bell I. Wiley, *Life of Johnny Reb,* Indianapolis & N.Y., 1943, p. 292; Bell I. Wiley, *Life of Billy Yank,* 1952, rep. Baton Rouge, La., 1978, p. 63.

Maynard Carbine

Dr. Edward Maynard, a dental surgeon, developed the uncomplicated breech-loading Maynard carbine. More than just a dental surgeon, Doctor Maynard also had a West Point education and an abiding interest in weapons. He invented and patented several gun parts in the 1840s and in 1845 invented a priming mechanism that saw wide use in both American military and civilian firearms. Then, in May 1856, he patented a breech-loading carbine.

Maynard's Carbine.

No. 361.

The Maynard appeared fragile compared to such heavy carbines as the Sharps or Springfield. Yet its simple, uncluttered design proved to be sturdy and, just as importantly, dependable. The Maynard became a favorite with Southern sportsmen and militia units just prior to the war.

The .50-caliber Maynard carbine weighed about six pounds and was 36 7/8 inches in total length. When the trigger guard was lowered, the barrel dropped, much like a shotgun. This allowed the loading of a special brass cartridge case with a hole to permit ignition. The hinged barrel was then locked back together at the breech. Using Maynard's patented roll of percussion caps, which were engaged on the nipple one at a time by cocking the hammer, the carbine was ready to fire. The roll of percussion caps worked similar to caps in the later toy cap pistol. The Maynard could be fired an average of ten rounds per minute. Only the fastest riflemen could fire a muzzle-loading musket as much as three times a minute.

One advantage of the Maynard design was that the cartridge case was of thick brass and could be reloaded and refired up to 100 times. The carbine also had excellent accuracy and range. In U.S. military testing in the late 1850s, the Maynard compiled a score of 41 hits out of 43 shots at 1,300 yards at a 10 x 30-foot panel. In spite of the distance, the Maynard's bullet still went completely through an inch of pine board. (At 500 yards, 80 percent of 250 rounds hit a 4-foot square canvas target.)

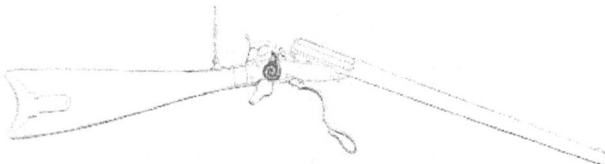

No. 865.
Maynard's Carbine ready for charge.

As the war drew near, increasing numbers of Maynards went South from Doctor Maynard Massachusetts factory to arm units mustering in the future Confederacy. As late as August 5, 1861, four months after the war had officially begun in Charleston Harbor, the firm of Sparks and Gallagher, in Louisville, Ky., sold 5,000 Maynards to the Confederates for $30 each.

"The Maynard rifle," according to one Confederate cavalryman quoted in 1864, "is the favorite with us, and proves a destructive weapon when one becomes accustomed to handling it, mounted, in a skirmish. It is light, simple in structure, and can be used with both caps; the only objection is you have to be careful in preserving the empty brass tubes, or you will not be able to make new cartridges..."

Shortly after the Battle of Sacramento, Bedford Forrest again showed his expertise with the Maynard. At Ft. Donelson, Tenn., in February 1862, Union sharpshooters were taking a toll of Confederates in their trenches and forcing the Southerners to keep their heads down or risk being shot. While inspecting the lines with his staff, Forrest saw a puff of smoke rise from the opposing treeline as a sharpshooter claimed another victim. He asked an aide for the loan of his Maynard. Quickly aiming, Forrest fired and the Federal fell dead from his perch. Forrest eventually came to prefer the greater power and reach of the muzzle-loading Enfield. But the little Maynard carbine helped him win many victories during the war.

Following the war, lever-action weapons such as the Henry—later the Winchester—would become the fashionable shoulder weapon. But the Maynard made a comeback as a sporting weapon. By 1873 it had again become a favorite with sportsmen, especially adaptable to a telescope sight.

Sources: Wayne Austerman, "Maynard," *Civil War Times Illustrated,* April 1986, pp. 42-45; Mark M. Boatner III, *The Civil War Dictionary,* 4th printing, N.Y., 1966; John S. Bowman, exec. ed., *The Civil War Almanac,* N.Y., 1983, pp. 287-288; Francis A. Lord, *Civil War Collector's Encyclopedia,* N.Y., 1965, pp. 237-240; John D. McAulay, *Carbines of the Civil War,* Union City, Tenn., 1981, pp. 51-55; Bell I. Wiley, *Life of Johnny Reb,* Indianapolis & N.Y., 1943, p. 292.

Sharps Carbine and Rifle
Christian Sharps of Philadelphia patented one of the first successful breech-loading shoulder arms in the U.S. in 1848. Counting those bought prior to 1861, more than 100,000 Sharps arms are estimated to have been used during the Civil War. The U.S. Government purchased 80,512 Sharps carbines and 9,141 Sharps rifles throughout the war period. The Confederates bought 1,600 of the rifles in February 1861. Many more were purchased by state authorities or by soldiers at their own expense, especially in the North. Each Sharps bought by the Federal government cost $36.15.

The Sharps relied on a vertical sliding breech block activated by a lever. The block was lowered and a paper cartridge inserted into the chamber. As the block was raised to close the breech, the sharp edge of the block sliced the paper from the cartridge, exposing the powder. On top of the block was a nipple for a percussion cap. Pulling the trigger dropped a hammer onto the cap; the resulting flash passed down a vent and struck the powder to fire the round. Sharps shoulder weapons were copied in the South, in a model known as "Richmond Sharps."

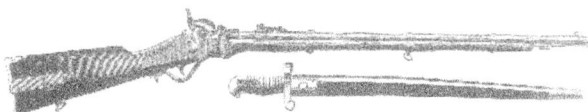

The Sharps was accurate to about 600 yards. A rate of up to 10 rounds a minute was possible, three times faster than a muzzle-loader could be fired. Weighing less than eight pounds, the .52-caliber Sharps carbine was 37 1/2 inches long. Its breech-loading made the Sharps carbine easier to load while on horseback, especially with its linen cartridge. The Sharps is considered the most popular single-shot breech-loading carbine of the war. Southern cavalrymen, especially, liked the Sharps carbine.

Sources: Mark M. Boatner III, *The Civil War Dictionary,* 4th printing, N.Y., 1966; John S. Bowman, exec. ed., *The Civil War Almanac,* N.Y., 1983, p. 288; Jack Coggins, *Arms & Equipment of the Civil War,* N.Y., 1962, pp. 34, 54; Francis A. Lord, *Civil War Collector's Encyclopedia,* N.Y., 1965, pp. 237, 253; John D. McAulay, *Carbines of the Civil War,* Union City, Tenn., 1981, pp. 13-21.

Shotguns
Most men in the South thoroughly understood the shotgun. It was part of everyday life in the rural South. Many of the horse soldiers not only preferred fighting with them, many did just that, especially early in the war. In fact, Confederate ordnance records show a surprising number of shotguns in use by Confederates all through the war.

On March 3, 1862, William Richardson Hunt, Southern ordnance officer in Memphis wrote Confederate Secretary of War J.P. Benjamin: "... Colonel Forrest, the most efficient cavalry officer in this department, informs me that the double-barrel shotgun is the best gun which cavalry can be armed." In fact, when Bedford Forrest published his request for enlistees in the Memphis, Tenn., *Daily Appeal,* he announced: "... I desire to enlist five hundred able-bodied men, mounted and equipped with such arms as they can procure (shot-guns and pistols preferable), suitable to the service..." While he had some troops throughout the war using shotguns, Forrest came to prefer the Enfield carbine, when he could capture them.

Although not as glamorous as a revolver, the muzzle-loading shotgun of the mid-nineteenth century had certain traits that recommended it for combat. In close combat, massed shotgun fire could be devastating. In the March 3 letter by William Hunt, he reported that Bedford Forrest had told him that at Fort Donelson one discharge of his cavalry's shotguns, at close quarters, had scattered 400 Federals that three Confederate regiments had been unable to dislodge from a ravine.

On the other hand, there were drawbacks. Most shotguns at that time were muzzle loaded, making it difficult to reload on horseback during a skirmish. Additionally, when a troop engaged Union cavalry in close combat, the scattering shot tended not to distinguish friend from foe. Its range also left much to be desired when Confederate cavalry armed with shotguns ran into Federals armed with rifles or carbines.

Sources: William A. Albaugh III and Edward N. Simmons, *Confederate Arms,* Harrisburg, Pa., 1957, p. 74; Jack Coggins, *Arms & Equipment of the Civil War,* N.Y., 1962, p. 54; Claud E. Fuller and Richard D. Steuart, *Firearms of the Confederacy,* Huntington, W. Va., 1944, pp. 209-210; Ashley Halsey Jr., *Who Fired The First Shot?,* N.Y., 1963, "What They Fought With," pp. 48-81, p. 51; Bell I. Wiley, *Life of Johnny Reb,* Indianapolis & N.Y., 1943, p. 292; John Allan Wyeth, *That Devil Forrest,* 1899, rep. Baton Rouge, La., 1989, pp. 21-22.

Battle of Sacramento Complete Bibliography
(includes some sources not directly quoted)
Books

Albaugh, William A. III, and Edward N. Simmons, *Confederate Arms,* Harrisburg, Pa., 1957.

Addington, Larry H. , *The Patterns Of War Since The Eighteenth Century,* Bloomington, Ind., 1984.

Allen, Felicity, *Jefferson Davis: Unconquerable Heart,* Columbia, Mo., 1999.

Anders, Curt, *Fighting Confederates,* NY, 1968.

Anderson, Charles R., *Vietnam: The Other War,* 1982, rep. N.Y., 1990.

Armann, William Frayne, ed., *Personnel of the Civil War,* N.Y., 1961.

Arthur, Anthony, *Bushmasters: America's Jungle Warriors of World War II,* 1987, rep. N.Y. 1989.

Ashdown, Paul, and Edward Caudill, *The Myth of Nathan Bedford Forrest,* N.Y., 2005.

Axelrod, Alan, *Patton On Leadership: Strategic Lessons For Corporate Warfare,* Paramus, N.J., 1999.

Bancroft, Hubert Howe, *The Works of Hubert Howe Bancroft,* Vol. 16: History of the North Mexican States and Texas: Vol. II 1801-1889, San Francisco, 1889.

Battle, J.H., "Todd County History," in J.H. Battle and W.H. Perrin, eds., *Counties of Todd and Christian, Kentucky,* Chicago and Louisville: 1884.

Bedwell, Randall, ed., *May I Quote You, General Forrest?,* Nashville, Tenn., 1997.

Bennett, Adrian A., *Missionary Journalist In China,* Athens, Ga., 1983.

The Biographical Encyclopaedia of Kentucky of the Dead and Living Men of the Nineteenth Century, Cincinnati, 1878.

Blackburn, John, *A Hundred Miles A Hundred Heartbreaks,* n.p., 1972.

Blair, Jayne E., *The Essential Civil War: A Handbook to the Battles, Armies, Navies and Commanders,* Jefferson, N.C., 2006

Blanton, Deanne, and Lauren M. Cook, *They Fought Like Demons: Women Soldiers in the Civil War,* 2002, rep. N.Y. 2003.

Blumenson, Martin, and James L. Stokesbury, *Masters of the Art of Command,* 1975, rep. N.Y., n.d.

Boatner, Mark M. III, *The Civil War Dictionary,* 4th printing, N.Y., 1966.

Bolin, James Duane, *An Abiding Faith: A Sesquicentennial History of Providence, Kentucky, 1840-1990,* Providence, Ky., 1990.

Bowman, John S., exec. ed., *The Civil War Almanac,* N.Y., 1983.

Bradford, Ernle, *Thermopylae: The Battle For the West,* 1980, rep. Cambridge, Mass., n.d.

Bradley, Michael R., *Nathan Bedford Forrest's Escort and Staff,* Gretna, La., 2006.

Brandon, Robert J., *Autobiography of Robert J. Brandon,* with additional notes by Edith Bennett, n.p., [1992].

Brandt, Nat, *The Man Who Tried To Burn New York,* Syracuse, N.Y., 1986.

Braverman, Libbie L., and Samuel M. Silver, *The Six-Day Warriors,* N.Y. 1969.

Breuer, William, *Devil Boats: The PT War Against Japan,* 1987, rep. N.Y., 1988.

Browning, Robert M. Jr., *Forrest: The Confederacy's Relentless Warrior,* Washington, D.C., 2004.

Bush, Bryan S., *Lloyd Tilghman: Confederate General in the Western Theatre,* Morley, Mo., 2006.

Calvert, Michael, with Peter Young, *A Dictionary of Battles: 1715-1815,* N.Y., 1979.

Carter, Howell, *A Cavalryman's Reminiscences of The Civil War,* 1900, rep. New Orleans, 1979.

Carter, Samuel III, *The Last Cavaliers: Confederate and Union Cavalry in the Civil War,* N.Y., 1979.

Carter, Gen. William H., *The U.S. Cavalry Horse,* 1895, rep. Guilford, Conn., 2003.

Castel, Albert, *The Guerrilla War 1861-1865,* Special Issue, *Civil War Times Illustrated,* Oct. 1974.

Cavalry Combat, U.S. Army Cavalry School, 1937.

Coggins, Jack, *Arms & Equipment of the Civil War,* N.Y., 1962.

Collins, Lewis, and Richard H. Collins, *History of Kentucky,* 2 vols., Covington, Ky., 1882, Vol. 1, "Annals of Kentucky," under date of Dec. 27.

Confederate Patriot Index: 1924-1978, Vol. II, Columbia, Tenn., 1978.

Connelly, Thomas Lawrence, *Army of the Heartland: The Army of Tennessee, 1861-1862,* Baton Rouge, La., 1967.

Cooke, Philip St. George, *Cavalry Tactics or Regulations for the Instruction, Formations, and Movements of the Cavalry of the Army and Volunteers of the United States,* 1862, rep Mechanicsburg, Pa., 2004.

Cotten, Michael, *The Williamson County Cavalry: A History of Company F, Fourth Tennessee Cavalry Regiment, C.S.A.,* n.p., 1994.

—————, *Williamson County Confederates,* n.p., 1996.

Cox, J.D., *March To The Sea,* 1883, rep. N.Y., 1963.

Coulter, E. Merton, *The Civil War and Readjustment in Kentucky,* Chapel Hill, N.C., 1926.

Creveld, Martin Van, *The Art of War: War and Military Thought,* 2000, rep. N.Y., 2005.

Crittenden, H.H., comp., *The Crittenden Memoirs,* N.Y., 1936.

Daniel, Larry J., *Days of Glory: The Army of the Cumberland 1861-1865,* Baton Rouge, La., 2004.

Dickson, Paul, *War Slang,* 2d ed., N.Y., 2000.

DiMarco, Louis A., *War Horse: A History of the Military Horse and Rider,* Yarkley, Pa., 2008.

Doyle, Don H., *Nashville in the New South: 1880-1930,* Knoxville, Tenn., 1985.

Drake, Edwin L., *Chronological Summary of Battles and Engagements of the Western Armies of the Confederate States,* Nashville, Tenn., 1879.

Duke, Basil W., *History of Morgan's Cavalry,* Cincinnati, 1868.

—————, *Reminiscences of General Basil W. Duke, C.S.A.,* N.Y., 1911.

Duncan, Thomas D., *The Recollections of Thomas D. Duncan,* 1922, rep., Tony Hays, ed. Savannah, Tenn., 2000.

Dupuy, Trevor N., Curt Johnson and David L. Bongard, *Harper Encyclopedia of*

Military Biography, Edison, N.J., 1992.

Dyer, Frederick H., *A Compendium of the War of the Rebellion,* Vol. III, Regimental Histories, 1908, rep. N.Y., 1959.

Estes, Claud, comp., *List of Field Officers, Regiments and Battalions in the Confederate States Army 1861-1865,* Macon, Ga., 1912.

Evans, David, *Sherman's Horsemen: Union Cavalry Operations in the Atlanta Campaign,* Bloomington, Ind., 1996.

Farwell, Byron, *Encyclopedia of Nineteenth-Century Land Warfare: An Illustrated World View*, N.Y., 2001.

Federal Writers' Project, *Military History of Kentucky,* Frankfort, 1939.

Feigel, Barbara Anne, *Civil War in the Western Ohio Valley As Viewed From Evansville, Indiana,* MA Thesis, Indiana University, Bloomington, Ind., 1957.

Foley, Charles, *Commando Extraordinary: Otto Skorzeny,* 1954, rep. London, Eng., 1998.

Ford, Roger, and Tim Ripley, *The Whites of Their Eyes: Close-Quarter Combat,* 1997, U.S.A. rep. Dulles, Va., 2001.

Fraser, Antonio, *The Warrior Queens,* 1988, rep. N.Y., 1989.

Fry, James B., *Killed by a Brother Soldier,* N.Y., 1885.

Fry, James C., *Combat Soldier,* Washington, D.C., 1968.

Fuller, Claud E., and Richard D. Steuart, *Firearms of the Confederacy,* Huntington, W. Va., 1944.

Fussell, Paul, *Wartime: Understanding and Behavior in the Second World War,* N.Y., 1989, paper rep., 1990.

Garber, Max B., and P.S. Bond, *A Modern Military Dictionary,* Washington, D.C., 1942.

Garrison, Webb, *A Treasury of Civil War Tales,* Nashville, Tenn., 1988.

----------, *Friendly Fire in the Civil War,* Nashville, Tenn., 1999.

----------, with Cheryl Garrison, *The Encyclopedia of Civil War Usage,* Nashville, Tenn., 2001

Gellhorn, Martha, *The Face Of War,* rev. ed., N.Y., 1988.

Gluckman, Arcadi, *United States Martial Pistols and Revolvers,* Buffalo, N.Y., 1944.

Gooch, J.T., ed., *Just The Other Day: A History of Madisonville, Kentucky,* Madisonville, Ky., 1981

Green, Thomas Marshall, *Historic Families of Kentucky,* 1889, rep. Baltimore, 1964.

Hafendorfer, Kenneth A., *Perryville: Battle for Kentucky,* Louisville, Ky., 1991.

Halsey, Ashley Jr., *Who Fired The First Shot?*, N.Y., 1963.

Hamilton, James, *The Battle of Fort Donelson,* N.Y., 1968.

Harlow, Alvin F., *Murders Not Quite Solved,* N.Y., 1938.

Harralson, Agnes S., *Steamboats on the Green,* Berea, Ky., 1981.

Harrison, Lowell H., *The Civil War in Kentucky,* Lexington, Ky., 1975.

Hays, William Foster, *Sixty Years of Owensboro: 1883-1943,* Owensboro, Ky., 1943.

Headley, John W., *Confederate Operations in Canada and New York,* N.Y., 1906.

Hemingway, Ernest, ed., *Men At War,* 1942, rep. N.Y., 1955.

Henderson, Harry McCorry, *Texas in the Confederacy,* San Antonio, Tex., 1955.

Henry, Robert Selph, *"First With the Most" Forrest,* 1944, rep. N.Y., 1991.

——————, ed., *as they saw Forrest,* Jackson, Tenn., 1956, pp. 33-35.

Herr, John K., and Edward S. Wallace, *The Story of the U.S. Cavalry 1775-1942,* Boston, Mass., 1953.

Hersey, John, *Of War And Men,* 1944, rep. N.Y., 1963.

History of Daviess County, Kentucky, 1883, rep. Evansville, Ind., 1966.

Hockersmith, L.D., *Morgan's Escape: A Thrilling Story Of War Times,* 1903, rep. n.p., n.d.

Hooker, Charles E., *Mississippi,* Vol. VII, *Confederate Military History,* Clement A. Evans, ed., Atlanta, Ga., 1899.

Hooker, Richard D. Jr., ed., *By Their Deeds: America's Combat Commanders On The Art Of War,* N.Y., 2003.

Horrall, S.F., *The History of the Forty-Second Indiana,* Chicago, Ill., 1892.

Hubbell, John T., and James W. Geary, eds., *Biographical Dictionary of the Union,* Westport, Conn., 1959.

Hurst, Jack, *Nathan Bedford Forrest: A Biography,* N.Y., 1993.

——————, *Men of Fire: Grant, Forrest, and the Campaign That Decided the Civil War,* N.Y., 2007.

Johnson, Adam R., *The Partisan Rangers of the Confederate States Army,* William J. Davis, ed., 1904, rep. Evansville, Ind., 1971.

Johnson, James Ralph, and Alfred Hoyt Bill (text); and Hirst Dillon Milhollen (illustration ed.), *Horsemen Blue and Gray: A Pictorial History,* N.Y., 1960.

Johnson, Robert Underwood, and Clarence Clough Buel, eds., *Battles and Leaders of the Civil War,* 4 Vols., 1887 - 1888, rep. N.Y., 1956, Vol. 3.

Johnson, Rossiter, *Campfire and Battlefield,* 1894, rep. N.Y., 1978.

Johnston, J. Stoddard, *Kentucky,* Vol. IX, *Confederate Military History,* Clement A. Evans, ed., Atlanta, Ga., 1899.

Johnston, William Preston, *The Life of Gen. Albert Sidney Johnston,* N.Y., 1879.

Jomini, Antonine Henri, *The Art of War,* 1862, rep. Westport, Conn., n.d.

Jordan, Thomas, and J.P. Pryor, *The Campaigns of Lieut.-Gen. N.B. Forrest, and of Forrest's Cavalry,* 1868, rep. Dayton, Ohio, 1973.

Keegan, John, and Richard Holmes, *Soldiers: A History of Men In Battle,* N.Y., 1985.

Keith, Elmer, *Six Guns,* N.Y., 1961.

Kentucky Adjutant General, *Report of the Adjutant General of Kentucky, Union, 1861-1866,* 2 vols., Frankfort, Ky., 1866-1867, Vol. 1.

——————, *Report of the Adjutant General of the State of Kentucky, Confederate,* Vol. II, Frankfort, Ky., 1918

Kinchen, Oscar A., *Confederate Operations in Canada and the North,* North Quincy, Mass., 1970.

Knapp, David Jr., *The Confederate Horsemen,* N.Y., 1966.

Kohn, George C., *Dictionary of Wars,* N.Y., 1986.

Leulliette, Pierre, *The War in Algeria: Memoirs of a Paratrooper,* 1961, rep. N.Y., 1987.

Lewis, Jon E., ed., *The Mammoth Book of True War Stories,* N.Y., 1992.

Lindsley, John Berrien, ed., *The Military Annals of Tennessee, Confederate,* Nashville, Tenn., 1886.

Lossing, Benson J., *Mathew Brady's Illustrated History of the Civil War* (originally, *A History of the Civil War),* 1912, rep. N.Y., n.d.

Lord, Francis A., *Civil War Collector's Encyclopedia,* N.Y., 1965.

Lytle, Andrew Nelson, *Bedford Forrest And His Critter Company,* N.Y., 1931.

Massey, Mary Elizabeth Massey, *Women in the Civil War,* originally *Bonnet Brigades,* 1966, rep. Lincoln, Neb., 1994.

Mauldin, Bill, *Up Front,* Cleveland, Oh., 1946.

McAulay, John D., *Carbines of the Civil War,* Union City, Tenn., 1981.

McDonough, James Lee, *War in Kentucky: From Shiloh to Perryville,* Knoxville, Tenn., 1994, paper rep. 1996.

McDowell, Robert Emmett, *City of Conflict,* Louisville, Ky., 1962.

McElroy, John, *Andersonville: A Story of Rebel Prisons,* 2 vols. , 1879, rep. Washington, D.C., 1899.

McLean, Wlliam E., *The 43d Regiment of Indiana Volunteers,* 1903, rep. Salem, Mass., 1998.

McLean Co., Ky., Will Book I.

McLean Co., Ky., Marriage Book A.

McLean County, Kentucky 1860 Census Annotated, Owensboro, Ky., 1978, Holly M. Leftwich, transcriber, Elizabeth S. Cox, annotater;

Medical and Surgical History of the War of the Rebellion, Joseph K. Barnes, ed., 6 vols., Washington, D.C., 1870-1888, Vol. 2, pt. 1, p. XL, pt. 2, p. 76 (listed as Dec. 28, 1862).

Meacham, Charles Mayfield, *A History of Christian County, Kentucky from Oxcart to Airplane,* Nashville, Tenn., 1930.

Meriwether, Lee, *My First 98 Years 1862-1960,* Columbia, Mo., 1960.

Mickle, William E., *Well Known Confederate Veterans And Their War Records,* New Orleans, La., 1907.

Military Operations of the Civil War: A Guide-Index to the Official Records of the Union and Confederate Armies, 1861-1865, Dallas Irvine, Edwin R. Coffee and Robert B. Matchette, comp., Washington, D.C., 1980.

Minor, Louisa H.A., *The Meriwethers and Their Connections,* Albany, N.Y., 1892.

Moore, Frank, ed., *The Rebellion Record: A Diary of American Events, with Documents, Narratives, Illustrative Incidents, Poetry, ETC.,* 11 vols., N.Y., 1861-1868, Vol. 3, Diary, Doc. 241; Vol. 4, p. 67.

Moore, Lt. Gen. (USA Ret.) Harold G., and Joseph L. Galloway, *We Are Soldiers Still,* N.Y. 2008.

Moore, James, *Kilpatrick and Our Cavalry,* N.Y., 1965.

Mundie, James A. Jr., Dean E. Letzring, Bruce S. Allardice and John H. Luckey, *Texas Burial Sites of Civil War Notables,* Hillsboro, Tex., 2002.

Muster, John W. III, personal interview.

Napoleon I, *The Military Maxims of Napoleon,* Lt. Gen. Sir George C. D'Aguilar,

David G. Chandler, commentary, 1831, rep. Mechanicsburg, Pa., 2002.

National Archives, General Services Administration, Washington, D.C., personal records.

Our Senior Soldiers: The Biographies and Autobiographies of Eighty Cumberland Presbyterian Preachers, Nashville, Tenn., 1915.

Perrin, W.H.; J.H. Battle and G.C. Kniffin, *Kentucky. A History of the State,* Louisville & Chicago, 1888.

Perrin, William Henry, ed., *Counties of Christian and Trigg,* Chicago & Louisville, 1884.

Phister, F., *Statistical Record,* N.Y., 1883.

Plutarch, *Plutarch On Sparta,* trans. with into and notes by Richard J.A. Talbert, N.Y., 1988.

Robinson, Charles M. III, *A Good Year To Die: The Story of the Great Sioux War,* N.Y., 1995.

Rommel, Erwin, *The Rommel Papers,* B.H. Liddell-Hart, ed., 1953, paper rep. N.Y., n.d.

Roscoe, Theordore, *Pig Boats* (originally published as *United States Submarien Operations in World War II*), rep. N.Y., 1958.

Rothert, Otto A., *A History of Muhlenberg County,* Louisville, Ky., 1913.

Saywell, Shelley, *Women In War,* N.Y., 1985.

Schiller, Laurence D., "A Taste of Northern Steel: The Evolution of Federal Cavalry Tactics 1861-1865," *North & South,* Vol. 2, No. 2 (Jan. 1999), pp. 30-45, 8084.

Scott, Douglas D; Richard A. Fox Jr.; Melissa A. Connor and Dick Harmon, *Archaeological Perspectives on the Battle of the Little Big Horn,* Norman, Okla., 1989.

Scott, Douglas D.; P. Willey and Melissa A. Connor, *They Died with Custer: Soldiers' Bones from the Battle of the Little Bighorn,* Norman, Okla., 1998.

Seven Early Churches of Nashville, Nashville, Tenn., 1972.

Shanklin, James Maynard, *"Dearest Lizzie": Civil War Letters of James Maynard Shanklin,* Kenneth P. McCutchan, ed., Evansville, Ind., 1988.

Sifakis, Stewart, *Who Was Who in the Union,* N.Y., 1988.

----------, *Compendium of the Confederate Armies: Tennessee,* N.Y., 1992.

----------, *Compendium of the Confederate Armies: Texas,* N.Y., 1995.

Simmons, Henry E., comp., *A Concise Encyclopedia of the Civil War,* N.Y., 1965.

Sipes, William B., *The Seventh Pennsylvania Veteran Volunteer Cavalry,* Pottsville, Pa., n.d.

Sistler, Barbara; Byron Sistler and Samuel Sistler, *1850 Census, North West Kentucky,* Nashville, Tenn. 1994.

Sklenar, Larry, *To Hell with Honor: Custer and the Little Bighorn,* Norman, Okla., 2000.

Speed, Thomas, *The Union Cause in Kentucky, 1860-1865,* N.Y., 1907.

Starling, Edmund L., *History of Henderson County, Kentucky,* Henderson, Ky., 1887.

Starnes, H. Gerald, *Forrest's Forgotten Horse Brigadier,* Bowie, Md., 1995.

Starr, Stephen Z., *The Union Cavalry in the Civil War:* Vol. I: *From Fort Sumter to Gettysburg 1861-1863,* Baton Rouge, La., 1979.

Steinbeck, John, *Once There Was A War,* 1958, rep. N.Y., 1960.

Steiner, Paul E., *Medical-Military Portraits of Union and Confederate Generals,* Philadelphia, 1968.

Stone, Richard G. Jr., *Kentucky Fighting Men 1861-1945,* Lexington, Ky., 1982.

Strode, Hudson, *Jefferson Davis American Patriot 1808-1861,* N.Y., 1955.

Sun Tzu, *The Art of War,* trans. with intro. by Marine Brig. Gen. Samuel B. Griffith (ret.), N.Y., 1963.

Tapp, Hambleton, and James C. Klotter, *Kentucky: Decades of Discord: 1865-1900,* Frankfort, Ky., 1977.

Taylor, William O., *With Custer on the Little Bighorn,* N.Y., 1996.

Tennesseans In The Civil War, 2 parts, Nashville, Tenn., 1964.

Tennessee: The Volunteer State: 1769-1923, Nashville, 1923, Vol. II.

3d Ky. Cav. Memo. Book, microfilm, National Archives

Thompson, Ed Porter, *History of the Orphan Brigade,* 1898, rep. Dayton, O., 1973.

Thucydides, *The Peloponnesian War,* c.400 B.C., rep. Rex Warner, trans., Baltimore, Md., 1974.

Truman, Arthur C. Jr., *At War With Ourselves 1861-1865,* Utica, Ky., 1992.

Turner, Josephine M., *The Courgeous Caroline Founder of the UDC,* Montgomery, Ala., 1965.

Union Soldiers and Sailors Monument Association, *The Union Regiments of Kentucky,* Louisville, 1897.

von Clausewitz, Carl, *On War,* Michael Howard and Peter Paret, eds. and trans., Princeton, N.Y., 1976, rep.ed with new essays, 1984.

von Luck, Hans, *Panzer Commander: Memoirs of Colonel Hans Von Luck,* 1989, rep. N.Y. 1991.

Wakelyn, Jon L., *Biographical Dictionary of the Confederacy,* Westport, Conn., 1977.

War of the Rebellion: A Compilation of the Official Records of the Union and Confederate Armies (commonly *Official Records of the Union and Confederate Armies*), 128 vols., index, and atlases, Washington, D.C., 1880-1901, Ser. I, Vol. 7, Vol. 49, Pt. 2; Ser. II, Vol. 3, Vol. 4.

Warner, Ezra J., *Generals In Gray: Lives of the Confederate Commanders,* Baton Rouge, La., 1959.

Wheeler, Joseph, *Alabama,* Vol. VII, *Confederate Military History,* Clement A. Evans, ed., Atlanta, Ga., 1899.

Wiley, Bell I., *Life of Johnny Reb,* Indianapolis & N.Y., 1943.

——————, *Life of Billy Yank,* 1952, rep. Baton Rouge, La., 1978.

Williams, Frances Marion, *The Story of Todd County, Kentucky,* 1820-1970, Nashville, 1972.

Williams, John, *World Atlas of Weapons & War,* London, Eng., 1976.

Wills, Brian Steele, *A Battle from the Start: The Life of Nathan Bedford Forrest,* N.Y., 1992.

Wittenberg, Eric J., "Learning the Hard Lessons of Logistics," *North & South,,* Vol. 2, No. 2 (Jan. 1999), pp. 62-69, 72-78.

Wood, C.E., *Mud: A Military History,* Washington, D.C., 2006.

Wooten, Dudley G., ed., *A Comprehensive History of Texas, 1685 to 1897,* Dallas, Tex., 1898.

Wright, Marcus J., *Texas in the War 1861-1865,* Harold B. Simpson, ed. and notes, Hillsboro, Tex., 1965.

Writers' Program, Works Project Administration, *Military History of Kentucky,* Frankfort, Ky., 1939.

Wyeth, John Allan, *Life of General Nathan Bedford Forrest,* N.Y., 1899.

Wyeth, John Allan, *That Devil Forrest,* 1899, rep. Baton Rouge, La., 1989.

Young, Bennett H., *Our Cavalry,,* n.p., n.d., booklet rep. from *Camp Fires of the Confederacy.*

Zhuge Liang and Liu Ji, *Mastering the Art of War,* Thomas Cleary, trans., Boston, Mass., 2000.

Articles and Correspondence

"A Brilliant Affair," *Memphis* (Tenn.) *Daily Appeal,* Jan. 2, 1862.

"An Affair At Sacramento," *Louisville* (Ky.) *Daily Journal,* Jan. 1, 1862.

Austerman, Wayne, "Maynard," *Civil War Times Illustrated,* April 1986, pp. 42-45.

Benton, Barbara, "Friendly Persuasion: Woman As War Icon, 1914-45," *MHQ: The Quarterly Journal of Military History,* Autumn 1983, pp. 80-87.

Campbell, D'Ann, "Women, Combat, and the Gender Line," *MHQ: The Quarterly Journal of Military History,* Autumn 1983, pp. 88-97.

Gilbert, Charles C., "On The Field of Perryville," in Robert Underwood Johnson and Clarence Clough Buel, eds., *Battles and Leaders of the Civil War,* 4 Vols., 1887 - 1888, rep. N.Y., 1956, Vol. 3, pp. 52-59.

"From Col. Hawkins's Regiment," *Louisville* (Ky.) *Daily Journal,* Feb. 10, 1862;

Gilliam, William D. Jr., "Family, Friends and Foe," *The Civil War in Kentucky,* supplement to *The* (Louisville, Ky.) *Courier-Journal,* November 1960, pp. 38, 41, 44.

"Girl on the Firing Line," *McLean Co. News,* Dec. 28, 1961.

Gray, H.T., "Forrest's First Cavalry Fight," *Confederate Veteran,* Vol. 15 (1907), p. 139.

Griese, Arthur A., "A Louisville Tragedy—1862," *Filson Club History Quarterly,* Vol. 26, No. 2 (April 1952), pp. 133-154.

Grimsley, Mark, "Millionaire Rebel Raider: The Life of Nathan Bedford Forrest," part 1, *Civil War Times Illustrated,* Oct. 1993, pp. 58-73.

Hancock, Gina, "Anniversary of battle is this week," quoting, among others, Cora Bennett, Elizabeth Cox, Martha Harris, Allen Taylor Nall and Kenny Ward, *McLean Co. News,* Dec. 31, 1987.

Harcourt, A..P., "Terry's Texas Rangers," *Southern Bivouac,* Vol. 1, No. 3 (Nov. 1882), pp. 89-97.

Haw, Joseph R., "The Last of C.S. Ordnance Department," *Confederate Veteran,*

Vol. 22 (1914), pp. 450-452.

Hill, Mrs. J.C., copier, "Cumberland Presbyterian Church Cemetery," in *McLean County, Kentucky Cemeteries, Volume I,* Owensboro, Ky., 1977, pp. 67-76.

"How Captain Bacon Was Killed," *Chicago* (Ill.) *Times,* Jan. 30, 1862.

"Jack O'Donnell," *Confederate Veteran,* Vol. 13 (1905), p. 175.

Johnston, D.W., "Correspondence," *Evansville* (Ind.) *Journal,* Jan. 1862.

Keating, J.M., "Tennessee For Four Years The Theater Of War," in John Berrien Lindsley, ed., *Military Annals of Tennessee. Confederate,* Nashville, Tenn., 1886, pp. 19-52.

Kelley, D.C., "Forrest's (Old) Regiment, Cavalry," in John Berrien Lindsley, ed., *Military Annals of Tennessee. Confederate,* Nashville, Tenn., 1886, pp. 761-769.

Kelly, C. Brian, "Badly wounded by his own men during the Civil War, 'Stovepipe' Johnson still would pursue his lifelong dream, *Military History,* April 2002, p. 82.

"The Late Capt. Bacon," *Louisville* (Ky.) *Daily Journal,* Jan. 3, 1862.

Lawrence, Keith, "Confederate day all but forgotten," Owensboro, Ky., *Messenger-Inquirer,* June 3, 1977.

Linden, Clarence, "Particulars of the Sacramento Skirmish," *Louisville* (Ky.) *Journal,* Feb. 24, 1862.

"Local Laconics," *Owensboro* (Ky.) *Messenger,* June 14, 1887.

Louisville (Ky.) *Democrat,* January 1862.

Louisville (Ky.) *Daily Journal,* January, February 1862.

Maslowski, Peter, "Reel War vs. Real War," *MHQ: The Quarterly Journal of Military History,* Vol. 10, No. 4 (Summer 1998), pp. 68-75.

Martin, Richard T., "Recollections of the Civil War," in Otto A. Rothert, *A History of Muhlenberg County,* Louisville, 1913, pp. 285-317.

Macon, Edna S., "Legacy of the Skirmish at Sacramento: the United Daughters of the Confederacy," *Kentucky's Civil War 1860-1865,* 2002-2003 edition, Clay City, Ky., 2002, p. 14.

Mangum, Neil C., "The General's Tour: The Little Bighorn Campaign: Civil War Veterans Die on the Plains," *Blue & Gray Magazine,* Vol. 23, No. 2, 2006, pp. 6-27, 42-65.

McCain, William D., "Nathan Bedford Forrest: An Evaluation," *Journal of Mississippi History,* Vol. 24, No. 4 (Oct. 1962), pp. 203-225.

McLean Co. News, Calhoun, Ky., Feb. 22, 1962; Dec. 31, 1987.

Memphis (Tenn.) *Daily Appeal,* Jan. 10, 1862.

"Military Affairs In Kentucky," *Western Citizen,* Paris, Ky., Jan. 10, 1862.

"More Prisoners," *Louisville* (Ky.) *Daily Journal,* January 11, 1862.

Morelock, Jerry D., "Hell To Pay: Bedford Forrest At Brice's Crossroads," *Armchair General,* November 2005, pp. 44-49.

"NEWS reader identifies heroine of Sacramento battle," quoting Cora Bennett, *McLean Co. News,* Feb. 22, 1962.

Niderost, Eric, "Mauled By The Russian Bear," *Military Heritage,* April 2006, pp. 26-33, 67.

Owensboro (Ky.) *Daily Messenger,* Oct. 7, 1894.

"Particulars of the Sacramento Skirmish," *Louisville* (Ky.) *Journal,* Feb. 24, 1862.

Protheroe, W.M., "It Was A Full Moon When... A Researcher's Guide to Civil War Moon Phases," *Blue & Gray Magazine,* July 1987, pp. 35-37.

"Rev. D.C. Kelley," *Confederate Veteran,* Vol. 17 (1909), p. 421.

Rhodes, Harry W., "Military Character of Gen. Forrest," *Confederate Veteran,* Vol. 3 (1895), p. 212.

Ridley, B.L., "Chat With Col. W.S. McLemore," *Confederate Veteran,* Vol. 8 (1900), p. 262.

Ridley, Bromfield L., "Chat With Col. W.S. McLemore," *Battles and Sketches of the Army of Tennessee,* Mexico, Mo., 1906, pp. 177-181.

Rikhoff, James C., "Guns of the Civil War: Part IV: Confederate Revolvers," *Gunsport,* Sept. 1960, pp., 34-36, 73-76.

Riser, Pearl J., "Nathan Bedford Forrest: 'The Wizard of the Saddle'," *United Daughters of the Confederacy Magazine,* April 1950, pp. 11-14.

Russell, Carl A., *Chronology of Reported Events, Gould's 23 Regiment,* "In memory of Thomas H. Massie, Private, Company B, Gould's Regiment, 23rd Texas Cavalry, C.S.A."

Sansing, David G., "Charles Clark Twenty-fourth Governor of Mississippi," *Mississippi History Now,* March 21, 2006 <http://mshistory.k12.ms.us/features/featured47/governors/19-charles-clark.html>.

Schonfeld, ... (Lt. Col), "Reconnaissance by Light Troops," trans. from *Wissen und Wehr,* 1941, in *The Cavalry Journal,* Vol. 50, No. 5 (Sept.-Oct. 1941), pp. 102-104.

Sheppard, Rev. T.J., "Religious Life and Work in Andersonville—How Captured...," in John McElroy, *Andersonville: A Story of Rebel Prisons,* 2 vols., 1879, rep. Washington, D.C., 1899, Vol. 2, pp. 628-638.

"The Skirmish At Sacramento, Ky.," *Chicago* (Ill.) *Times,* Jan. 6, 1862.

Smith, J. Robert, "Lt. Col. Robert Martin was talk of Gen. Forrest's army," *Times-Argus Messenger Magazine,* Central City, Ky., Sept. 11, 1969.

Southard, Thomas, "Sacramento sixth graders report more local Civil War stories," *McLean Co. News,* Calhoun, Ky., Nov. 27, 1958.

Stier, William J., "Fury Takes The Field," *Civil War Times Illustrated,* December 1999, pp. 40-48.

Stone, Henry Lane, "Reminiscences of Morgan's Men," *Southern Bivouac,* Vol. 1, No. 11 (July. 1883), pp. 406-414.

Tipps, H.T., "History of McKendree United Methodist Church, in *Seven Early Churches of Nashville,* Nashville, Tenn., 1972, pp. 7-21.

Todd, Ann Vance, Aug. 7, 1994, correspondence with author, containing typed copies of depositions of Isaac and Elizabeth Johnson.

Turner, Josephine, "Caroline Meriwether Goodlett: Founder, United Daughters of the Confederacy," *United Daughters of the Confederacy Magazine,* Sept. 1962, pp. 12-14, 36-41.

Utley, J. Harold, "The Battle of Sacramento... With Nathan Bedford Forrest & Madisonville's Al Fowler," Battle of Sacramento Planning Committee, *"Down Memory Lane" in Sacramento, Kentucky,* n.p., 1999.

"W.N.M.," "Sketch of Lieutenant-General N.B. Forrest," *Southern Bivouac,* Vol. 2, No. 7 (Mar. 1884), pp. 289-298.

"War Reminiscences," *Owensboro* (Ky.) *Messenger,* Jan. 29, 1888.

Ward, John K., "Skirmish at Sacramento: Battle of Future Generals," *Register* of the Kentucky Historical Society, Frankfort, Ky., Vol. 75, No. 2 (1977), pp. 79-91.

—————————, "Forrest's First Fight," *America's Civil War,* March 1993, pp. 50-57.

Weller, Joe, "Nathan Bedford Forrest: An Analysis of Untutored Military Genius," *Tennessee Historical Quarterly,* Vol. 18, No. 3 (Sept. 1959), pp. 213-251.

Williams, T. Harry, "Kentucky, The Hard School of Experience," *The Civil War in Kentucky,* supplement to *The* (Louisville, Ky.) *Courier-Journal,* November 1960, pp. 53-54, 58.

Winter, William F., "Mississippi's Civil War Governors," *Journal of Mississippi History,* Vol. LI, No. 2 (May 1989), pp. 77-88.

Wolseley, Garnet, "General Viscount Wolseley on Forrest," in Robert Selph Henry, *as they saw Forrest,* Jackson, Tenn., 1956, pp 17-53.

Woodworth, Steven E., *Jefferson Davis and His Generals: The Failure of Confederate Command in the West,* 1982, rep. Lawrence, Kan., 1990.

Wright, Marcus, "Memorandum of General Officers," in *Personnel of the Civil War,* Vol. 1, N.Y., 3d printing, 1968, pp. 207-376.

Line Art Sources

Page 11—from Robert Underwood Johnson and Clarence Clough Buel, eds., *Battles and Leaders of the Civil War,* 4 Vols., N.Y., 1887 - 1888, Vol. 1, p. 398.

Page 27—from Rossiter Johnson, *Campfire and Battlefield,* 1894, rep. N.Y., 1978, p. 197.

Pages 93-99—from *Shuyler, Hartley & Graham Illustrated Catalog of Civil War Military Goods,* 1864, rep. Dover Publications, Mineola, N.Y., 1985, pp. 127-142

Index

Page numbers in **Bold** denote pictorial representation.

Custer, Thomas, 62
Cypress Creek, 84

D

Daviess Co., Ky., 4, 19, 72
Davis, Arthur N., 16, 17, **17**, 21, 25, 63-64
Davis, Jefferson, 60, 61, 67, 82
Davis, Jefferson C., 73
Davis, Joseph, 61
Decherd, Tenn., 73
Denby, Charles, 74
Denton Co., Tex., 75
Department of the Ohio (U.S.), 90
Diabetes, 67, 86
Dibrell's Tennessee Brigade (C.S.), 82
Diodorus, 3(in boxed quotes)
District of Western Kentucky (C.S.), 76
Duke, Basil W., 35
Duke, Charlton G., 58, 59
Duke, John C., 58
Duncan, Thomas D., 36

E

8th Kentucky Cavalry (U.S.), 69
8th Tennessee Cavalry Battalion (C.S.), 3, 5, 10, 82, 87
El Salvador, 12
11th Kentucky Infantry (U.S.), 4, 7
Enfield Rifles, 10, 24, 94-95, **94**, 96-97, 98
English, 17, 18, 29
Evans, John W., 25
Evansville, Ind., 80

F

Falkland War, 32
Fall, Albert Boult, 57
Fayette Co., Ky., 8
5th Division, Army of the Ohio (U.S.), 6
5th Virginia Cavalry (C.S.), 89
1st Cavalry Division (Vietnam War), 4 (in boxed quote)
1st Division, IX Corps (U.S.), 61
1st Kentucky Cavalry (Mexican War), 8
1st Kentucky Cavalry (C.S.), 4, 59, 83
1st Louisiana Cavalry (C.S.), 57
1st Missouri Infantry (C.S.), 59
Fishburg (?Frostburg), 10
Foertsch, Herman, 23 (in boxed quote)
Forrest, Jesse, 66
Forrest, Mary, 67
Forrest, N.B. Jr., 67
Forrest, Nathan Bedford, numerous mention throughout, **2, 64**
Forrest's Old Regiment (C.S.), 64
Ft. Donelson, Tenn., 1, 3, 64, 69, 70, 88, 96, 98
Ft. Henry, Tenn., 1

Ft. Pillow, Tenn., 65
Ft. Pulaski, Ga., 60
42d Indiana Infantry (U.S.), 10, 55, 74
42d Virginia Infantry (C.S.), 63
48th Tennessee Infantry (C.S.), 88
Fourteenth Brigade, 5th Division (U.S.), 55
4th Tennessee Cavalry (C.S.), 82, 87
Fowler, Alvin 'Al', 120, 68-70, **69**
Frankfort, Ky., 22, 56, 57, 79
Frankfort (Ky.) Home Guards, 57
Franklin Co., Ky., 56
Franklin Tenn., 82
Frederick the Great, 15, 44
Frostburg, 10
Fry, James, 33

G

Gainesville, Tex., 75
Galloway, Joseph L., 24 (in boxed quote), 25 (in boxed quote)
Galt House, 9
Gardiner, Lee M., 80
Garst's Pond, **10**, 11
Gellhorn, Martha, 18, 24 (in boxed quote)
Germans, 29
Goodlett, Michael Campbell, 83
Gould, Nicholas C., 2, 13, **13**, 28, 36, 70-71
Grant, U.S., 65
Gray, T.H., 12
Greece, 1
Greeks, 27
Green River, 1, 5, 6, 9, 18, 30, 57, 76, 81
Greenup Co., Ky., 9
Greenville, Ky. 5, 6, 8, 20, 21, 24, 25, 30, 57, 68, 90, 91
Grubb's Cross Roads, Ky., 76
Gwin, Larry, 25 (in boxed quote)

H

Hackett, Rowland E., 59
Haley's Frankfort (Ky.) Brass Band, 57
Halley, ?, 59
Halpeth River, 4
Hammer, James, 12, 24, 28
Hampton, Thomas W., 28
Hannibal, 16
Harris, Isham G., 2
Harris, J. R., 88
Hatley, Sherwood, 59
Headley, John, 4, 35
Hedrix, Reagan, 27, 28
Helm, Benjamin Hardin, 83
Hemingway, Ernest, 13, 14 (in boxed quote)
Henderson, Ky, 6, 7

Ripley, J.W., 20
Roberts, John J., 40
Robinson, R.R., 19
Rochester, Ky., 4, 40
Rommel, Erwin, 12 (in boxed quote)
Rosecran, William S., 60
Rothert, Otto, 45
Rumsey, 4, 5, 8, 39
Russellville, Ky., 4, 62

S

Sabers(sabres), 17, 19, 21, 22, 23, 44, 48, 63
Sacramento, Ky., (numerous mentions throughout)
Sacramento, Ky., Cumberland Presbyterian Church, 21, 47
St. Louis (Mo.) Republic, 53
Salisbury, N.C., POW camp, 44
San Diego, Calif., 60
Savage pistols, 65
Saywell, Shelley, 9 (in boxed quote), 9
Scott, Gus A., 40
Scott, John A., 40
2d Brigade, Forrest's Division (C.S.), 61
2d Mississippi Volunteers (Mexican War), 42
Selma, Ala., 46
17th Infantry (post war), 43
7th Cavalry Regiment (post war), 44
Shacklett, J.O., 21
Shanklin, James M., 8
Sharps, Christian, 68
Sharps rifle, carbine, 11, 67, 68
Sheppard, T.J., 18
Sheridan, Phil, 12
Sherman, William T., 8, 11, 60
Shiloh, Tenn., Battle of, 31, 60, 61, 64, 70, 73, 76, 85
Shotguns, 13, 15, 44, 66, 68-69
Sidell, W.W., 7
Sims, M.L., 49
Singleton, William, 57
16th Illinois Cavalry (U.S.), 18
Smith, A.J., 45
Smith, E. Kirby, 50
South Carrollton, Ky., 4, 8, 39
Spanish Civil War, 22 (in boxed quote)
Sparks and Gallagher, 67
Sparta, 13
Springfield Rifle, 66. 67
Staked Plains Station, Tex., 5
Stanton, Shelby L., 4 (in boxed quote)
Starling, Samuel M., 52
Starnes, James W., 3, 4, 6, 8, 10, 11, 16, 19, 20,, **22**, 23, 26, 61-62
Staten Island, N.Y., 43

20th Infantry Regiment (post-war), 44
26th Kentucky Infantry (U.S.), 8, 21, 42
26th Tennessee Cavalry Battalion (C.S.), 25
23d Texas Cavalry (C.S.), 49, 50
Typhoid Fever, 58

U

Underwood, Uriah M., 17
United Daughters of the Confederacy (UDC), 58
University of Misssissippi, 43
Utah, 60

V

Vanderbilt University, 54, 55, 58
Veteran Reserve Corps (U.S.), 42
Vicksburg, Miss., 49
Vietnam War, 20, 23

W

Wade, Daniel F., 61
Wallace, B.P., 41
Walters, John L., 16, 17, 18, 62-63
Wardlaw, Caroline, 56
Ware, E.M., 62
Waugh, Evelyn, 13
Webster, George, 52
Webster Co., Ky., 48
Weir Farm, 6
Wheeler, J.B., 58
White, Josiah, 2
Whitmer, Catherine, 59
Whitmer, Daniel, 59
Whitworth Rifle, 62
Wiggins, Q.K. Juniper, 39
Williams, John L., 15, 17, 18, 48, 63-64
Williams, Joseph M., 58
Williamson Co., Tenn., 4, 57
Williamson County Cavalry (C.S.A.), 3, 4
Wilson, Ewing A., 4
Wilson Co., 57
World War I, 6
World War II, 6, 9, 12, 13, 15, 21, 23, 25, 47
Wyeth, John Allan, 57

X

Xerxes, 15, 19, 25

Y

Yaffa, 9 (in boxed quote)

Z

Zorndorf, Battle of, 15, 44

John Kenneth (Ken) Ward

John Kenneth (Ken) Ward was born in McLean Co., Ky., within five miles of the end of the fight at Sacramento. Of pioneer stock, his area ancestry dates back to at least 1790. He graduated from Calhoun High School, Calhoun, Ky., and subsequently entered Lockyears Business College, Evansville, Ind., and worked in Evansville before enlisting in the U.S. Army. His six years military service included more than two years in Vietnam. Following military service and working on the Ohio River as a deckhand for the Corps of Engineers, Ward attended the University of Kentucky. Here he received a Bachelor of Arts in History and has since taken several graduate-level courses in specific areas of history.

A free-lance writer for more than 40 years, he has published more than 125 historical articles. These include articles published in *America's Civil War, Military History, Register of the Kentucky Historical Society, American Heritage of Invention & Technology, Kentucky Living, Early American Life,* etc. His published article, "Ancient Warrior's Revenge," was chosen for inclusion in the anthology *Best of the Wild West,* published by the National Historical Society. He wrote a historical column for *The McLean Co. News,* Calhoun, Ky., for more than 30 years, and during 1981-1982 also wrote weekly historical articles for *The Jessamine Times,* Nicholasville, Ky.

Ward has worked as a historical consultant for two consulting firms on projects in the Southeast and Upper South. At the request of the late Harry W. Caudill, Ward wrote a section on the Civil War in Eastern Kentucky for a projected historical survey of Appalachia. As a guest lecturer, he has spoken at schools in Calhoun and Lexington and a Lexington retirement community. A member of the Civil War Preservation Trust and the Army Historical Foundation, he is an honorary member of the Sons of Confederate Veterans. He is also a Founding Sponsor of the National Museum of the U.S. Army.

Presently residing in Lexington, Ky., he works as a patient relations associate in the Emergency Room, Chandler Medical Center, University of Kentucky. And he still indulges in his love of researching and writing.

John K. (Ken) Ward (standing) and his brother, Hugh, in a 19th century fantasy.

www.ingramcontent.com/pod-product-compliance
Lightning Source LLC
Chambersburg PA
CBHW060400090426
42734CB00011B/2202